BURLEIGH DODDS SCIENCE: INSTANT INSIGHTS

NUMBER 59

Bacterial diseases affecting pigs

burleigh dodds
SCIENCE PUBLISHING

Published by Burleigh Dodds Science Publishing Limited
82 High Street, Sawston, Cambridge CB22 3HJ, UK
www.bdspublishing.com

Burleigh Dodds Science Publishing, 1518 Walnut Street, Suite 900, Philadelphia, PA 19102-3406, USA

First published 2022 by Burleigh Dodds Science Publishing Limited
© Burleigh Dodds Science Publishing, 2022. All rights reserved.

British Library Cataloguing in Publication Data
A catalogue record for this book is available from the British Library

ISBN 978-1-80146-414-7 (Print)
ISBN 978-1-80146-415-4 (ePub)

DOI: 10.19103/9781801464154

Typeset by Deanta Global Publishing Services, Dublin, Ireland

Contents

Series list

Chapter 1

Diseases affecting pigs: an overview of common bacterial, viral and parasitic pathogens of pigs

Alejandro Ramirez, Iowa State University, USA

1 Introduction

Diseases affecting pigs can be quite complex. It is well recognized that often these conditions are multifactorial, especially as in the case of respiratory diseases, hence the term 'porcine respiratory disease complex' (PRDC) has been accepted (Brockeier, 2002). The concept of the disease triad (Fig. 1) emphasizes this complexity and the interaction between not only different pathogens, but the host, pathogen and the environment.

To maintain a sustainable pork production system, we must move away from the idea of one agent-one disease and look at the whole picture from a holistic point of view. Recent world events concerning pigs such as the 2009 pandemic influenza outbreak, the continuous spread of African swine fever virus (ASFv) in Eastern Europe and the introduction of several new pathogens into the United States (porcine epidemic diarrhoea virus and porcine deltacoronavirus) and their spread to Canada (limited), Mexico, Central and South America have

http://dx.doi.org/10.19103/AS.2017.0013.14

Disease Triad

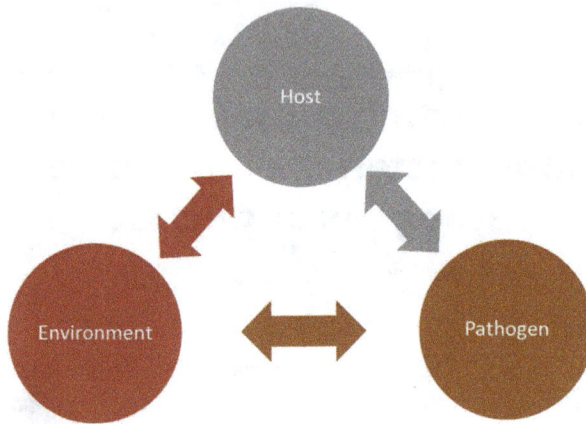

Figure 1 Depiction of the disease triad demonstrating the interactions between host, pathogen and the environment.

emphasized that we live in one world with minimal borders. Diseases of pigs spread rapidly across various countries.

The concept of one world and one health requires all of us to work together to improve the health and welfare of pigs to achieve sustainable pork production.

The following section summarizes some of the most common bacteria, viruses and parasites found in pig production including those of greatest importance due to their effect on production, importance from an international trade perspective, as well as their zoonotic concern. It is not to be considered a comprehensive review by any means.

2 The most common bacterial pathogens in pig production: gram-negative bacteria

The advent of new diagnostic technologies such as polymerase chain reaction (PCR) testing have allowed pig veterinarians to diagnose a wide range of viral diseases, making bacterial infection appear to be 'old' pathogens. However, bacterial pathogens continue to significantly affect the health and well-being of pigs, and pig bacterial pathogens are of great importance today because of human concern regarding antimicrobial resistance. Worldwide, pig farmers and veterinarians are being pressured to use less antimicrobials stressing the importance of their responsible and judicious use and disease prevention instead.

2.1 *Actinobacillus pleuropneumoniae*

Actinobacillus pleuropneumoniae (APP) is a gram-negative coccobacillus and has global importance as the causative agent of pleuropneumonia in pig. Pleuropneumonia is characterized as a highly contagious disease with sudden onset, high mobility and mortality, only affecting pigs. Many different serotypes are recognized based on the RTX exotoxins secreted by the organism. Currently, four of these exotoxins are recognized (ApxI, ApxII, ApxIII, ApxIV) and their presence varies among serotypes (Chiers et al., 2010). These toxins cause severe internal pulmonary haemorrhage and cytotoxicity. Apart from these toxins, many other virulence factors have also been identified. The virulence of serotypes, as well (asymptomatic to high mortality) as their prevalence in different geographical regions, is highly variable (Clota et al., 1996; Mittal et al., 1996; Kucerova et al., 2005; Gottschalk, 2003). It is well recognized that isolates from most herds have more than one low-virulent serotype (Gottschalk, 2003).

One of the major challenges with APP is that pigs continue to carry the organism in their lungs and tonsils for several months (Desrosiers, 2004), thus creating an opportunity for repeated outbreaks of this clinical disease, as well as creating a challenging environment at all stages of pig production, from growing pigs until slaughter. Antimicrobials are used to treat the disease with several countries, especially the United States, demonstrating a pattern of resistance to beta-lactams (penicillin, amoxicillin). Antimicrobial treatment helps minimize mortalities during the early stages of an outbreak, but they do not eliminate carrier pigs and Sjölund et al. (2009) have suggested that the use of highly effective antimicrobials prevents good antibody response to the infection, thus leaving pigs susceptible to future re-infections.

Vaccination against APP is challenging due to the various different serotypes. They involve stimulation by several different Apx exotoxins as well as an outer membrane protein (Gottschalk, 2012).

The severity of APP infections is a significant burden on the sustainability and feasibility of an infected herd. This is especially true today with the current emphasis on antimicrobial stewardship, thus infected herds must undergo a de-population and re-population many times with APP-free pigs.

There are no zoonotic concerns regarding APP other than it can lead to extensive antimicrobial use in infected herds.

2.2 *Bordetella bronchiseptica*

Bordetella bronchiseptica is a gram-negative rod that is found throughout the world and infects many different mammalian species (Brockmeier et al., 2012). In pigs *Bordetella bronchiseptica* primarily causes pneumonia and atrophic rhinitis.

Several virulence genes which appear to require co-expression of the BvgAS genes are identified (Beier and Gross, 2008) and are also subject to phase variation. Early BvgAS genes are involved in bacterial attachment followed later (once a large number of bacteria have colonized the area) with gene expression for toxin production (Brockmeier et al., 2012). It is this toxin production (especially dermonecrotic toxin) that contributes to the progression of the disease, especially atrophic rhinitis (nasal turbinate and septal damage).

Vaccination of sows before farrowing can be used in conjunction with antimicrobials to prevent atrophic rhinitis in pigs. The aim of these interventions is to prevent or minimize early colonization by *Bordetella bronchiseptica*, allowing the pig's immune system to protect it as it matures as well as preventing the late stages of BvgAS gene expression, which produces the dermonecrotic toxin. Vaccination helps minimize disease but does not prevent infection. Antibiotics can help minimize disease transmission between pigs.

Bordetella bronchiseptica often works in conjunction with toxigenic *Pasteurella multocida* to cause the more severe disease known as progressive atrophic rhinitis (de Jong and Nielsen, 1990).

Although human infections can occur with *Bordetella bronchiseptica* they are rare, and pigs do not appear to be a concern for zoonosis.

2.3 *Brachyspira spp.*

The *Brachyspira hyodysenteriae* is a spirochetal bacterium that causes swine dysentery (SD). Herds infected with SD incur serious financial losses due to mucoid and bloody diarrhoea resulting in a significant number of deaths as well as poor growth performance.

SD affects the large intestine of grower and finisher pigs but rarely affects weaners (Hampson, 2012). With the advent of PCR technology and genetic sequencing, several new species of *Brachyspira* have been identified and shown to cause SD lesions in growing pigs (Burrough et al., 2012; Chander et al., 2012). So although technically SD is only associated with *Brachyspira hyodysenteriae*, today the phenotypic culture characteristics (especially hemolysis) of *Brachyspira* spp. appear to be a more sensitive indicator of potential to induce dysentery-like disease in pigs (Table 1) than molecular identification alone (Burrough et al., 2012).

There is a limited arsenal of antimicrobials which can be used to treat SD. Mice and rats can serve as an important reservoir for *Brachyspira hyodysenteriae* (and maybe other *Brachyspira* spp.), which make it difficult to completely eliminate an infection from a herd. The limited weapons available are expensive antimicrobials that can add a significant cost to SD control. Therefore, de-population and re-population with SD-free pigs may be necessary in conjunction with aggressive cleaning, disinfection of premises and extensive

Table 1 Clinical significance of some *Brachyspira* spp. in pigs (Burrough 2015, pers. comm.)

	Hemolysis	Clinical disease in pigs
Brachyspira hyodysenteriae	Strong beta-hemolysis	Swine dysentery
Brachyspira hampsonii	Strong beta-hemolysis	Dysentery-like disease
Brachyspira suanatina	Strong beta-hemolysis	Dysentery-like disease
Brachyspira pilosicoli	Weak beta-hemolysis	Spirochetal colitis (mild disease)
Brachyspira murdochii	Weak beta-hemolysis	Mild to non-pathogenic
Brachyspira intermedia	Weak beta-hemolysis	Mild to non-pathogenic
Brachyspira innocens	Weak beta-hemolysis	Non-pathogenic

rodent control programmes. Currently there are no effective vaccines against *Brachyspira* spp.

There are no zoonotic concerns regarding *Brachyspira hyodysenteriae* other than there may need to be extensive antimicrobial use in infected herds.

2.4 *Brucella suis*

Brucella suis is a gram-negative coccobacillus of zoonotic importance in pigs, usually causing reproductive failure including long-term, non-fatal, granulomatous inflammation in several organs (including joints and reproductive organs such as testes). Its impact on production is mostly noted in acute outbreaks as frequently endemic brucellosis is mild enough that it can go undetected. Feral pigs are often found to be infected with *Brucella suis* throughout the United States (Leuenberger et al., 2007; Leiser et al., 2013).

The long duration of bacteraemia (5 weeks to 34 months) reported by Deyoe (1967 and 1972) suggests that immune response to infection is insufficient to eliminate bacteria from blood and intracellular niche, especially in macrophages (Olsen et al., 2012). Treatment or vaccination is not recommended because of the zoonotic potential of this organism, and the animal should be removed from the herd.

Although human brucellosis is extremely important, swine brucellosis is of lesser importance since milk from pigs is rarely consumed (Pappas et al., 2006). The zoonotic potential for infection due to prolonged bacteraemia and the relatively low infectious dose for humans is especially important due to work-related exposure and wild boar hunters.

2.5 *Escherichia coli*

Escherichia coli is a gram-negative aerobic rod that can be associated with many diseases in pigs including diarrhoea, septicaemia, mastitis and urinary tract infections. Often with pigs the different *E. coli* are identified by their fimbrial

Table 2 Most common fimbrial types for *Escherichia coli* in pigs and age susceptibility

Neonatal	Post weaning
F4 (K88)	F4 (K88)
F5 (K99)	F18
F6 (987P)	
F41	

type which vary depending on the age of the pigs affected (Table 2). Vaccination can be effective if the correct fimbrial type is included in the vaccine. It is of particular interest to note that pigs under 20 days of age are significantly less susceptible to F18 *E. coli* due to the lack of receptors at this age.

Most *E. coli* are considered zoonotic prompting a food safety concern, although it is important to note that pigs are not considered a normal source for enterohaemorrhagic *E. coli* O157:H7 (Fairbrother and Nadeau, 2006). Table 3 lists some of the common enterotoxins associated with *E. coli* in pigs.

2.6 *Haemophilus parasuis*

Haemophilus parasuis (HPS) is a gram-negative bacterium which causes Glässer's disease (fibrinous polyserositis) with at least 15 different serovar groups identified (Kielstein and Rapp-Gabrielson, 1992). There does not appear to be a direct association between serovars and virulence. One of the challenges with HPS is that it tends to be more common in high health status herds. HPS often appears after co-mingling (mixing) pigs from different sources (Wiseman et al., 1989). In acute outbreaks, HPS tends to affect the biggest and healthiest pigs in the group. HPS invades endothelial cells and causes apoptosis and production of pro-inflammatory cytokines creating accumulation of fibrin (Vanier et al., 2006; Bouchet et al., 2008).

Strategic use of antimicrobials can be used to mitigate the sudden effects of HPS by administering it at times of high risk, especially during stressful events (weaning and animal movement). Unfortunately, after killing the bacteria, the antimicrobials do nothing to the fibrin already produced. As this fibrin dries,

Table 3 Enterotoxins associated with *Escherichia coli* in pigs

Enterotoxin	Name	Effect
STa	Heat-stable toxin A	Decreases absorption of water and electrolytes
STb	Heat-stable toxin B	Increase fluid secretion by enterocytes
LT	Heat-liable toxin	Increase secretion of Na$^+$, Cl$^-$ and HCO$_3^-$
Stx2e	Shiga-like toxin	Increase vascular permeability

it becomes fibrous and can affect heart and lung movements, resulting in a chronically ill pig. Effectiveness of HPS vaccination can be variable (Oliveira et al., 2004; Oh et al., 2013).

There are no zoonotic concerns regarding HPS.

2.7 Lawsonia intracellularis

Lawsonia intracellularis is an obligate intracellular bacterium which grows preferentially in intestinal epithelial cells, causing ileitis in pigs. There are three quite distinct clinical presentations of the disease:

1 the traditional porcine intestinal adenomatosis (thickening of intestine);

2 the more chronic form of enteritis with fibrinonecrotic membrane; and

3 the peracute haemorrhagic form resulting in sudden death.

The importance of this disease is its continuous effect on decreasing feed efficiency and average daily gain (McOrist et al., 1997; McOrist, 2005). The peracute form usually affects market-ready pigs, resulting in sudden death with significant consequential economic losses.

Prevention involves the use of vaccines and/or strategic pulsing with a variety of antimicrobials (McOrist et al., 1999; Hammer, 2004; Bak and Rathkjen, 2009).

There are no zoonotic concerns regarding *L. intracellularis* infection in pigs.

2.8 Pasteurella multocida

Pasteurella multocida is a gram-negative rod or coccobacillus which can cause pneumonia and atrophic rhinitis in pigs. Most pig isolates are either serotype A or D, whereas serotypes B, C and E are found in cattle, reindeer and water buffalo. Most *P. multocida* Type A have a predilection to lung tissue while Type D are usually involved in progressive atrophic rhinitis along with *Bordetella bronchiseptica*, although either type can be found in the other's preferred tissue (Carter, 1955; Pijoan et al., 1983; Rimler and Rhoads, 1987).

P. multocida is the most common bacterial infection found in PRDC and it is the primary target for antimicrobial therapy. There are toxigenic and non-toxigenic strains of *P. multocida*. It is interesting to note that *P. multocida* by itself cannot cause pneumonia even when heavily inoculated. This suggests that there is a need for a primary co-infection to enable the establishment of *P. multocida* (Brockmeier et al., 2001). Vaccines containing *P. multocida* toxin have been effective against progressive atrophic rhinitis in pigs. *P. multocida* has no food safety concerns but has the potential for being zoonotic.

2.9 *Salmonella spp.*

Salmonella spp. are gram-negative rods known to infect a broad range of hosts. The two most important salmonella in pigs are *S. choleraesuis* (pigs only) and *S. typhimurium* (humans and pigs).

Clinical signs of salmonellosis can be variable in pigs depending on the strain. Both *S. choleraesuis* and *S. typhimurium* cause diarrhoea in pigs while *S. choleraesuis* more often tends to be systemic causing cyanosis of the skin as well as an interstitial pneumonia (Schwartz, 1997; Foley and Lynne, 2008; Carlson et al., 2012). There are over 200 virulence factors that have been identified with *Salmonella*, but few have been fully characterized (Carlson et al., 2012).

Vaccination is quite effective in helping prevent disease and antimicrobial use helps with treatment. As a gram-negative enteric pathogen, drug resistance (via plasmids) is common (Barnes and Sorensen, 1975; Schwartz, 1997). In addition to vaccination and antimicrobial therapy, bio-security with heavy emphasis on cleanliness to minimize faecal-oral exposure is important.

Salmonella is of zoonotic concern impacting on food safety.

3 The most common bacterial pathogens in pig production: gram-positive bacteria

3.1 *Clostridium spp.*

Clostridium spp. are anaerobic gram-positive spore-forming rods with several different species causing different diseases in pigs. *Clostridium* spp. cause disease via the different toxins they produce. In this chapter we will only discuss the two *Clostridium* spp. of greatest concern in pigs.

3.1.1 *Clostridium difficile*

In pigs, the clinical signs usually appear in the first few hours or days of life. The disease is believed to be caused by two toxins (Toxins A and B) and the administration of equine-origin antitoxins can mitigate the effects (Ramirez et al., 2014). Antimicrobial use does not appear to affect the severity of the disease in neonatal pigs (Arruda et al., 2013), which makes sense as the microflora is barely established in newborn piglets immediately after birth. Interestingly, although *C. difficile* infections are not seen in older pigs (more than 7 days of age), work by Arruda et al. (2013) has shown pigs are still susceptible at 10 days of age. There are currently no effective pig vaccines against *C. difficile*, which can be found in the faeces of most mammals.

In humans, *C. difficile* infections can be very serious or even deadly with antimicrobial-associated diarrhoea (Bartlet et al., 1978). In humans it can result in simple diarrhoea, colitis, pseudomembranous colitis, ileus, toxic mega colon and even bowel perforation (Kelly et al., 1984). However, there is currently no data to directly link *C. difficile* infection in pigs to zoonotic issues.

3.1.2 Clostridium perfringens

There are currently five different toxinotypes (Table 4), with only toxinotypes A and C affecting pigs. Enteritis by *C. perfringens* Type C has been well characterized and pre-farrowing vaccination programmes have been effective in controlling this disease quite well. On the other hand, *C. perfringens* Type A is still a bit of an enigma. Traditionally, Type A infections have only been associated with alpha toxin production. Unfortunately, as seen in Table 3, all other *C. perfringens* toxinotypes also produce this same toxin. Several research labs have suggested a role for a beta2 toxin with this disease (Bueschel et al., 2003; Waters et al., 2003), while others more recently question its role (Faranz et al., 2013). Field vaccination with *C. perfringens* Type A toxoid does not appear to be as effective as Type C vaccination.

There are no zoonotic concerns regarding *C. perfringens* in pigs. Human food poisoning with *C. perfringens* is mostly associated with consumption of beef, poultry or gravies.

3.2 Tuberculosis

Tuberculosis continues to be responsible for significant economic losses for pig producers in many countries while others such as the United States have practically eliminated it from their pig population (Thoen, 2012). Pigs are susceptible to *Mycobacterium avium* complex (MAC) and *M. tuberculosis* complex amongst other mycobacterial species (Thoen et al., 1975).

Table 4 Five *Clostridium perfringens* toxinotypes and their respective toxin and animals they can infect

Clostridium perfringens	Animals affected						Toxins				
	Pigs	Sheep	Goats	Poultry	Cattle	Horses	Alpha	Beta	Beta2	Epsilon	Iota
Type A	X	X		X	X		X		X		
Type B		X			X	X	X	X		X	
Type C	X	X	X	X	X	X	X	X			
Type D		X	X		X		X			X	
Type E		?			X		X				X

Pigs often acquire MAC when reared on ground contaminated by poultry (Schalk et al., 1935) and sometimes even sawdust (Schliesser and Weber, 1973). Most cases in pigs are asymptomatic or non-specific and therefore only diagnosed at slaughter (Thoen, 2012). Slaughter inspection specifically looks for granulomatous lesions in lungs or lymph nodes (granulomatous lymphadenitis). However, current European Union guidelines on pig meat inspection discourage the palpation/incision of such post-mortem lesions during routine slaughter in an effort to minimize bacterial cross-contamination (EUFSA, 2011)

As with most diseases that cause granulomas, the use of antimicrobials is not recommended due to long duration of treatment and poor prognosis. There are no effective vaccines available for pigs.

Although MAC can be a significant zoonotic and food safety risk for humans, especially those with immunocompromised immune systems (e.g., elderly, AIDS), pigs and pork have not been implicated as an exposure risk for human infection (Arasteh et al., 2000; Thoen, 2012).

3.3 *Mycoplasma spp.*

Mycoplasmas are a type of bacteria that lack cell walls. There are several mycoplasmas of importance in pigs including *Mycoplasma hyopneumoniae* (Mhyop), *M. hyorhinis* (Mhyor), *M. hyosynoviae* (Mhyos) and *M. suis*. Because of the large number of different mycoplasmas, it is important to be specific in name and not just refer to them as simply 'mycoplasma'.

There are no zoonotic concerns regarding any of the *Mycoplasmas* in pigs.

3.3.1 *Mycoplasma hyopneumoniae*

Mhyop is the aetiologic agent for enzootic pneumonia in pigs, one of the most significant bacterial respiratory pig pathogens worldwide. The strains of Mhyop are antigenically diverse (Frey et al., 1992; Thacker and Minion, 2012). Mhyop is difficult to grow in most laboratories.

Transmission of Mhyop occurs via nose-to-nose contact especially from sow to pig (Calsamiglia and Pijoan, 2000) but can also occur via aerosol up to 3.2 km (Goodwin, 1985) and 9.2 km (Otke et al., 2010). The organism attaches to ciliary epithelium of the respiratory tract and grows slowly. Protein P97 is involved in adherence (Zhang et al., 1994). Mhyop also alters the function of macrophages (Caruso and Ross, 1990) as well as other parts of the immune system (Thacker and Minion, 2012). Mhyop is an important potentiate of other respiratory pathogens in association with PRDC.

Vaccination of growing pigs can be considered 'standard' in today's pig production. Antimicrobials can also be used strategically to mitigate Mhyop as well as other bacterial co-infections.

3.3.2 *Mycoplasma hyorhinis*

Mhyor is associated with polyserositis and arthritis in three- to ten-week-old pigs. Current interest in this pathogen has increased due to welfare concerns of lameness in growing pigs. Little is known regarding the virulence and pathogenesis of Mhyor (Thacker and Minion, 2012). With the advent of newer PCR technology, many samples are now being tested for Mhyor, possibly creating a false sense of increased prevalence in recent years. Bacterial culture for Mhyor can be quite easy but requires special media (Ross and Whittlestone, 1983).

There are several antimicrobials which are used to treat Mhyor infection, although efficacy is quite variable (Thacker and Minion, 2012). It is suspected that part of the problem is late diagnosis of the disease. As with any joint infection, early detection is key.

3.3.3 *Mycoplasma hyosynoviae*

Mhyos is mostly associated with arthritis, and is very similar to Mhyor except that Mhyos tends to affect older pigs (3-5 months of age). As is the case with Mhyor, the Mhyos bacteria can be found in the tonsils of infected and 'normal' pigs (Thacker and Minion, 2012). Bacterial culture for Mhyos can be quite straightforward but requires special media (Friis et al., 1992).

3.3.4 *Mycoplasma suis*

Eperythrozoon suis, now renamed *Mycoplasma suis*, infects red blood cells of pigs causing anaemia (moderate to severe), respiratory distress (Doyle, 1932) and possible reproductive problems. The organism can live in the cytoplasm as well as in membrane-bound vacuoles of erythrocytes (Groebel et al., 2009) which can make it inaccessible to many antimicrobials. Bacterial culture for *M suis* is not yet possible so diagnosis is currently done via PCR.

3.4 *Staphylococcus spp.*

Staphylococcus spp. are gram-positive cocci that are regarded as normal bacterial flora of adult pig skin (Frana, 2012). There are two primary *Staphylococcus* of importance in pigs: *S. hyicus* and methicillin-resistant *S. aureus* (MRSA).

3.4.1 *Staphylococcus aureus*

Staphylococcus aureus can often be isolated from the skin of pigs as well as from septicaemia, mastitis, metritis and metritis infections. Although *S. aureus* rarely causes disease, recent attention to a specific type of *S. aureus* known as MRSA has stimulated interest in this bacteria. In particular, a unique MRSA known as ST398 was first associated with pigs in Europe (Armad-Leferve et al., 2005). This same sequence type does not appear to be as prevalent or important in pig production in the United States (Frana et al., 2013).

MRSA appears to be asymptomatic in pigs and is not considered to be a herd problem. Although on-farm antimicrobial use is suspected in MRSA, no studies have been able to demonstrate this association, which brings into question the ethical legitimacy of using stigmatization as a direct means to achieve public health outcomes (Plough et al., 2015).

Although MRSA is a significant human health concern, outside Denmark, the role of pigs in MRSA zoonosis does not appear to be significant. MRSA is not a food safety concern.

3.4.2 *Staphylococcus hyicus*

Staphylococcus hyicus is the causative agent for greasy pig disease or exudative epidermitis. This condition has worldwide distribution and presents as a skin infection (pyoderma) in young pigs (nursery age or younger). Although *S. hyicus* is commonly found in pig skin, under the right conditions the bacterium will establish itself in the epidermis via an abrasion in the skin. In severe cases the loss of fluids and electrolytes can lead to dehydration and death. There are several exfoliative toxins that have been identified and are considered the most important virulence factors for greasy pig disease (Amtsberg, 1979).

High humidity in pens, as well as a high number of young gilts farrowing, contributes to higher incidence or acute outbreaks of the disease. Injectable antimicrobials are commonly used to treat affected pigs along with topical treatments (spraying/soaking) which may involve the use of disinfectants (Frana, 2012). Pig farmers often use autogenous vaccines against *S. hyicus* with variable effectiveness.

There are no zoonotic concerns regarding *S. hyicus* in pigs.

3.5 *Streptococcus suis*

Although there are many other *Streptococcus* spp. that infect pigs, we will only discuss *S. suis*. *Streptococcus suis* is a gram-positive encapsulated coccus that can frequently be found in the tonsils, intestines and genital tract of healthy pigs (Gottschalk, 2012). There are 35 serotypes based on capsular polysaccharide but serotype 2 is the most common and most important one in pigs because of its zoonotic potential.

There are a large number of virulence factors identified for *S. suis* but none are fully understood due to the complexity of multiple factors (Braums and Valentin-Weigand, 2009). *Streptococcus suis* infection in pigs can be variable, in part due to variations in virulence and may include septicaemia, central nervous signs (meningitis), arthritis, pneumonia, vegetative valvular endocarditis, rhinitis and abortions (Sanford and Tilker, 1982).

Vaccination against *S. suis* is often ineffective. Beta-lactams and macrolide antimicrobials are commonly used to prevent and treat *S. suis* infections.

In South-East Asia, *S. suis* is the most common cause of bacterial meningitis in humans and therefore has been identified as a serious emerging public health threat (Wertheim, 2009). *S. suis* poses a zoonotic concern including a food safety concern in countries with particular cultural practices and preferences such as drinking uncooked blood from infected pigs and eating organs such as the uterus.

4 The most common viral pathogens in pig production

As a general rule it is helpful to know if the viral pathogen is a DNA or an RNA virus as well as whether it is an enveloped or non-enveloped virus. Compared with DNA viruses, RNA viruses mutate often as they do not have the necessary proofing mechanism when they replicate. Thus there can be great variability between strains and developing vaccines can be more challenging. Non-enveloped viruses tend to be much more resistant to inactivation than enveloped viruses. This means that non-enveloped viruses tend to persist longer in the environment. Clearly there are always exceptions to the rule, but these guiding principles can be very useful when learning about a new virus and its possible behaviour regarding transmission between pigs. Many of these important viruses are listed by the World Organization for Animal Health (OIE) as notifiable diseases due to their importance regarding animal and human health as well as international trade.

4.1 African swine fever virus

ASFv is an enveloped DNA virus from the family *Asfarvideae* that causes a highly contagious and haemorrhagic disease in pigs of all ages. ASF is an OIE-listed disease with important international trade consequences. The disease is endemic in sub-Saharan Africa and has now spread to several Eastern European countries. There are many different hosts for the virus as well as significant variation in virulence between strains (De Villeret al., 2010).

ASFv can be transmitted by many means including soft ticks and direct contact with contaminated oral and nasal secretions (Colgrove et al., 1969), consumption of contaminated feed and possibly short distances via aerosol (CFSPH, 2015a). The virus can survive long periods in cured meats (Mebus et al., 1993). ASFv can infect multiple tissues but their primary cells for replication include monocytes and macrophages (Malmquist and Hays, 1960; Minguez et al., 1988). Although ASFv does not induce neutralizing antibodies (De Boer, 1967), protective immunity against homologous but not heterologous re-infection still occurs (Ruiz Gonzalvo et al., 1981).

Currently there are no good vaccines against ASFv and treatment is not recommended as disease eradication should be the goal.

ASFv is not zoonotic but the severity of the disease causes significant food security and sustainability concerns.

4.2 Aujeszky's disease virus

Aujeszky's disease, also known as pseudorabies, is caused by the herpes virus, which is an enveloped DNA virus whose only natural host is pigs. Aujeszky's disease presents with central nervous system indicators, reproductive problems including abortions, respiratory illness and mortality. All other mammals, except humans, are end hosts for the virus, resulting in close to 100% mortality in these species. Its distribution is worldwide, with variations in virulence between strains and is an OIE-reportable disease.

Aujeszky's disease virus is primarily transmitted between pigs via direct and indirect contact including long-distance aerosol (Christen et al., 1990). Even after recovery, pigs remain infected for the rest of their life (common amongst herpes virus infections) and stress can reactivate viral shedding, helping to spread the disease to other pigs (Wittmann and Rziha, 1989).

Gene-deleted vaccines can be used with DIVA (differentiate vaccinated from infected) capabilities. They are quite effective at protecting against viraemia and clinical signs but unfortunately they do not prevent latent infections. Aujeszky's disease is not of zoonotic concern.

4.3 Classical swine fever virus

Classical swine fever (CSF), also known as hog cholera, is caused by an enveloped RNA Pestivirus that causes generalized systemic disease indistinguishable from many other common, endemic bacterial and viral pig diseases. CSF is an OIE-reportable disease. CSF is endemic in parts of Asia, South and Central America and some Caribbean islands (CFSPH, 2015b).

CSF virus is highly contagious and can cause septicaemia, anorexia, constipation, diarrhoea, lethargy and abortions, to mention just a few. The variable clinical signs and virulence of the virus is dependent on many factors including variation in strains (asymptomatic to high mortality) (Depner et al., 1997; Moenning et al., 2003).

As with natural infections, a combination of cell-mediated immunity and neutralizing antibodies appear to be important in producing sterilizing immunity (Pirou et al., 2003). There are several vaccines, especially live or modified live, that provide good protection against disease including some oral vaccines for wild boars (CFSPH, 2015b). CSF disease is not of zoonotic concern.

4.4 Coronaviruses

There are several different coronaviruses of importance in pigs. For the most part, they have quite similar clinical presentation, primarily causing diarrhoea with similar treatment and control but immunologically are very distinct from each other (i.e. no cross-protection between these different viruses). They are enveloped RNA viruses with spike proteins that give them a crown-like appearance under an electron microscope. None of the pig coronaviruses are of zoonotic concern.

4.4.1 Deltacoronavirus

With the introduction of porcine epidemic diarrhoea virus (PEDv) into the United States in early 2013 a new porcine deltacoronavirus (PDCov) was identified in faeces from pigs with diarrhoea (Wang et al., 2014). Work done by June et al. (2015) demonstrated that this virus is capable of causing disease in gnotobiotic pigs. Anecdotal field evidence suggests PDCov can cause diarrhoea, but is significantly milder and shorter in duration than PEDv. Currently there are no vaccines for PDCov.

4.4.2 Porcine epidemic diarrhoea virus

The importance of PEDv has re-surfaced after its introduction into the United States in May 2013. This introduction highlighted the possible role feed and feed ingredients can play in disease transmission – an area previously ignored most of the time. The clinical presentation for the disease is almost exactly the same as that of transmissible gastroenteritis virus (TGEv). The challenge of any new disease introduced into a naïve population was quite apparent. Disease spread easily from place to place and the infectious dose of the virus was quite low as data to be published in peer-review journals will show. Recent research suggest faecal viral shedding occurs within one day of exposure, peaks around 7 days and can continue for over 28 days (Magstadt et al., 2014). Interestingly, faecal consistency in the same study was clinically normal at 10 days post-inoculation, even though PEDv shedding was still occurring.

Clinical disease is more severe in piglets less than 3 weeks of age, with piglets less than 10 days old usually experiencing 100% mortality in an acute outbreak. Clinical signs can be short in duration to asymptomatic in older pigs. Immunity to PEDv appears to be short term (probably less than 4 months) and both colostral and lactogenic immunity are important in protecting baby pigs (Thomas, 2014). Vaccines can be helpful in herds with previous field exposure to PEDv but do not appear to be as effective when used in naïve animals (Thomas, 2014; Schelkopf et al., 2015). Although currently there are at least three different reported isolates in the United States, research suggests there is still good cross-protection between these different isolates (Zhang et al., 2015). Current vaccines against PEDv appear to be effective when used in animals previously exposed to live virus (Jung and Saif, 2015).

4.4.3 Transmissible gastroenteritis virus

The occurrence of TGEv has significantly decreased over the past decade as the appearance and widespread prevalence of the TGEv respiratory mutant (porcine respiratory coronavirus) spread throughout the United States suggesting possible cross-protection between these two related pathogens (Yaeger et al., 2002).

Clinical presentation, treatment and control options are the same for TGEv and PEDv. Vaccines against TGEv are not very effective and produce partial immunity of short duration (Saif et al., 1994).

4.5 Foot-and-mouth disease virus

Foot-and-mouth disease virus (FMDv) is a member of the *Picornaviridae*, a family of non-enveloped RNA viruses of significant international importance. It appears that FMDv is one of the best recognized and dreaded livestock diseases causing severe vesicular lesions in cloven-hooved animals and is at the top of the OIE-reportable diseases. FMDv is endemic in large areas of Africa, Asia, the Middle East and South America.

Seven serotypes (O, A, C, SAT 1, SAT 2, SAT 3 and Asia 1) which are important for vaccination are recognized. Serotype O is the most common serotype worldwide (CFSPH, 2014). Foot-and-mouth disease (FMD) can be transmitted via direct, indirect and aerosol means, with all secretions and excretions from infected animals containing the infectious virus (Alexandersen et al., 2012). Unlike ruminants, pigs do not become carriers or harbour FMDv for more than 28 days. FMDv is quite resistant and can remain infectious in the environment, cured meats and dairy products for several weeks (Bachrach, 1968; Cottral, 1969; CFSPH, 2014).

Vesicular lesions from FMD are clinically indistinguishable from any other vesicular disease including swine vesicular disease, vesicular exanthema of swine, vesicular stomatitis and Seneca virus A. Mortality in adults tends to be low, but animals have difficulty eating and moving around, resulting in a welfare concern.

Vaccination requires matching the proper serotype as there is no cross-protection between all seven serotypes. Most FMD vaccination has been focused on cattle and not pigs. There is also a wide range of strains within each serotype, thus complicating vaccine efficacy (Kitching et al., 1989). This requirement to match the strain and serotype of FMDv makes it difficult to react to an emergency outbreak. Vaccine-induced protection is short, lasting only about 4–6 months necessitating at least two doses per year (Domenech et al., 2010). Most of the largest producers are free of FMD and do not vaccinate.

The first OIE/Food and Agriculture Organization of the United Nations Global Conference on FMD led to the development of a Global FMD Control Strategy. This effort is focused on improving FMD control in regions where the disease is still endemic through the use of a Progressive Control Pathways (PCP) tool and is supported by many countries including the European Commission for the Control of Foot-and-Mouth Disease (EUFMD). This PCP offers a structured five-stage approach to FMD control, allowing FMD-endemic countries to become more successful in achieving FMD-free status strategically (OIE and FAO, 2012).

FMDv is not zoonotic, but the severity of the disease and international trade restrictions as an OIE-listed disease causes significant food security and sustainability concerns.

4.6 Influenza A virus in swine

Influenza A viruses in swine (IAv-S) are members of the family *Orthomyxoviridae*, which are enveloped viruses with segmented RNA, and can cause respiratory infections in most mammals. The segmented genome of IAv-S facilitates the exchange of gene segments between different IAv-S which may infect the same cell. This rearrangement of genes can generate new strains of the virus. Virus replication is limited to the upper and lower respiratory tract (Van Reeth et al., 2012).

The primary IAv-S are H1N1 and H3N2. Within each of these influenza types there are many different groupings of strains and cross-protection between strains, even within one type, can be quite variable. Infection occurs primarily via direct nose-to-nose contact between pigs and between people and pigs but more rarely between pigs and people (Nelson, 2014).

Vaccination continues to be the best means for prevention (Van Reeth et al., 2012). With such large diversity in field strains, matching the correct vaccine isolates to field isolates is extremely challenging, especially in livestock where vaccine regulations limit how quickly vaccine isolates can be changed. It is important to note that maternal antibodies do interfere with vaccine efficacy.

IAv-S are of zoonotic importance but are not a food safety concern. The biggest public health fear is the rearrangement of IAv-S into a novel strain infecting a naïve human population as was the case in the 2009 influenza pandemic.

4.7 Porcine circovirus type 2 virus

Porcine circovirus 2 viruses (PCV2v) belong to the family *Circoviridae* and are non-enveloped DNA viruses of global importance in pigs. PCV2v causes a variety of systemic diseases in pigs including wasting, pneumonia, late-term abortions, stillbirths, porcine dermatitis, nephropathy syndrome and diarrhoea (Harding and Clark, 1997).

PCV2v causes immunosuppression in pigs, making them vulnerable to a wide variety of infections (Chianini et al., 2003). The characteristic case definition for post-weaning multi-wasting syndrome (PMWS) requires three components:

1 lymphoid depletion,

2 large number of PCV2v in the lesion, and

3 clinical signs of wasting with a doubling of mortality (Sorden, 2000).

PCV2-infected animals develop good neutralizing antibodies in 10–28 days (Pogtanichniy et al., 2000; Fort et al., 2007). Vaccines are extremely effective in preventing PCV2-associated disease and is a standard part of any vaccination programme for growing pigs.

PCV2v is not of zoonotic concern, but the severity of the disease causes significant food security and sustainability concerns.

4.8 Porcine reproductive and respiratory syndrome virus

Porcine reproductive and respiratory syndrome virus (PRRSv) is an enveloped RNA virus from the family *Arteriviridae*, which affects pigs. PRRSv is present in most pig-producing regions, although there are a few countries with significant pig production where the virus has never been documented. Porcine reproductive and respiratory syndrome (PRRS) is an OIE-reportable disease.

One of the most significant characteristics of PRRSv is its ability to mutate. PRRSv has been reported to have a mutation rate multiple times that of the human immunodeficiency virus. This high mutation rate results in the development of quasi-species (a grouping of more than one genetic sequence related to a common mutation in an animal at the same time) (Rowland et al., 1999; Goldberg et al., 2003). This quasi-species is important because when genetically sequencing a sample from an infected animal, it is actually obtaining the consensus sequence of the different PRRSv viruses present. The high mutation rate also makes it difficult to find a single vaccine isolate that will provide broad cross-protection.

PRRSv infects macrophages, especially pulmonary macrophages (Thanawongnuwech et al., 1997; Duan et al., 1997). Targeting pulmonary macrophages results in an increased susceptibility to secondary infections in pigs. Protective immunity against PRRSv is slow to develop (4–6 weeks), so frequently requires closure of the herd for around 200+ days (Torremorell et al., 2002). The slow protective immunity along with low transmissibility of the virus has generated sub-populations of animals with different levels of immunity (Dee et al., 1996).

There are many vaccines available in the market with none able to provide universal protection against all strains. With the diversity of PRRSv, cross-protection becomes difficult to predict. Even today with sequencing technology, this genetic information only helps with epidemiological investigations but cannot be used in any way to predict cross-protection. New technology has identified genetic resistance in some pigs (Rowland et al., 2012) as well as the creation in 2015 of the first genetically engineered pigs completely resistant to PRRSv (Basi, 2015).

PRRSv is not of zoonotic concern, but the severity of the disease causes significant food security and sustainability concerns.

4.9 Rotaviruses

Rotaviruses are non-enveloped RNA viruses of the family *Reoviridae* and are a major cause of diarrhoea in neonatal and young pigs. There are four different porcine rotavirus serogroups identified (A, B, C and E). It was not until the last five to ten years that PCR technology has enabled the detection of serogroups other than type A (Médici et al., 2011). This new technology has resulted in increased detection and awareness of rotavirus B and C as well as the concept of co-infections with more than one rotavirus at a time.

Rotaviruses are highly prevalent throughout the world with some countries demonstrating up to 100% sero-prevalence in adult pigs (Chang et al., 2012).

Rotaviruses replicate predominately in villous epithelium in the small intestine (Buller and Moxley, 1988) as well as the large intestine (Theil et al., 1978) causing villous atrophy and diarrhoea. The high prevalence of rotavirus in the field suggests piglets are constantly being exposed to the virus and are likely to have reduced performance due to infections. Rotavirus infections are well recognized in neonatal pigs but are often ignored post-weaning.

Colostrum and lactogenic immunity play an important role in helping protect neonatal piglets from clinical disease (Saif, 1999; Wagstrom et al., 2000). Currently there are only vaccines against rotavirus type A infections.

Porcine rotaviruses are not considered to be of zoonotic concern.

5 The most common parasitic pathogens in pig production

Modern pig production has moved pigs indoors and thus limited their exposure to many internal and external parasites. Unfortunately, there is a general misconception that indoor pigs do not have internal parasites. Although many indoor facilities are clean and disinfected between groups of pigs, eggs from internal parasites are quite resistant. Personal experience suggests that the discontinued use of anthelmintic in indoor production has allowed some internal parasites to slowly propagate, increasing the exposure dose of growing and breeding pigs.

5.1 Ascarids

There are several ascarids that can infect pigs but *Ascaris suum* (pig round worm) is the most common and most important one. The life cycle is direct, taking four weeks for eggs passed in faeces to develop infectivity. These ascarid eggs are highly resistant to desiccation allowing them to build up in outdoor as well as indoor facilities (Barrett, 1976).

As infective eggs are consumed by the pig, the larvae will hatch in the intestine and travel through the intestinal wall to the liver, causing the traditional milk spots (scars) The larvae then travel on to the lungs causing verminous pneumonia before being re-swallowed completing the life cycle by developing into adults in the small intestine. This whole migration process causes internal damage reducing pig growth and production. *Ascaris suum* are zoonotic but are not a food safety concern.

5.2 Cysticerci

The adult tapeworm (*Taenia solium*) produces larvae which are then ingested by pigs and hatch into cysticercus (*Cysticercus cellulosae*). They are found in the skeletal and cardiac muscle of infected pigs and are referred to as 'measly pork' or 'pork measles'.

Pigs do not appear to have many clinical signs when infected but they can serve as a source of infection for humans (Greve, 2012). Pigs cannot re-infect themselves, infection requires ingestion of human tapeworm eggs. Pigs must not be allowed to ingest human faeces.

Cysticercosis is not a concern for modern pig production farms due to bio-security practices. In 2010 the World Health Organization added cysticercosis as a neglected tropical disease of zoonotic food safety concern, especially in undeveloped countries with free-roaming pigs (WHO, 2016). In humans cysticerci have a predilection for the central nervous system, making this a serious disease.

5.3 Coccidia

Coccidia are obligate intracellular protozoan with two main coccidia in pigs. Pigs can be infected with both *Isospora suis* or *Eimeria* spp.

5.3.1 Isospora suis

Isospora suis is the primary protozoal disease of neonatal pigs while *Eimeria* spp. are rarely identified. There appears to be a seasonal incidence of this disease as warmer temperatures and higher humidity favours sporulation of *I. suis* (Stuart and Lindsay, 1986). *I. suis* is not affected by the use of traditional coccidiostats as employed in other livestock (decoquinate, Amprolium, sulphas and ionophores). Prevention is focused on increased sanitation.

I. suis is not zoonotic.

5.3.2 Eimeria spp.

Eimeria spp. can infect pigs but are rarely identified in them (Lindsay et al., 1987). There are many different species of *Eimeria* which can be found in pigs worldwide. Although clinical disease in pigs is rare there have been sporadic reports of clinical diarrhoea in pigs of different ages (Hill et al., 1985). There are no studies on treatment or control options for *Eimeria* spp. Prevention should be focused on increased sanitation. *Eimeria* spp. are not zoonotic.

5.4 Sarcoptes scabiei

Sarcoptic mange (*Sarcoptes scabiei*) is the most important external parasite of pigs globally as the mite creates a highly pruritic condition affecting average daily gain, feed efficiency and even reproductive performance in herds (Kessler et al., 2003). The intense itching causes significant property damage with consequential financial losses for the pig farmer.

Eradication programmes can be effective because *S. scabiei* lives and completes its entire life cycle on the skin of pigs and environmental contamination is fairly trivial (Smith, 1986). Mange is not zoonotic or of food safety concern.

5.5 Trichuris suis

The pig whip worm *Trichuris suis* occurs primarily in the cecum of pigs and can cause diarrhoea with or without blood and mucous, affecting growth rate and feed efficiency. Its clinical significance is that many anthelmintics used for roundworms are ineffective against whipworms.

5.6 Trichinella spiralis

Trichinae in pigs is usually caused by *Trichinella spiralis*, which have minimal effect on pigs, but have significant health effects on people. Garbage feeding, as well as pig access to infected rodent carcasses or other dead pigs, is the primary means for transmission. Raising pigs indoors with limited access to wildlife along with aggressive rodent control has practically eliminated this disease from commercial pigs in the United States (Greve, 2012). This elimination of Trichinae has also allowed for new lower cooking recommendations for pork in the United Sates (from 71°C to 63°C), enabling people to enjoy a tastier (less dry) pork chop (USDA, 2011). *Trichinella spiralis* is of great zoonotic and food safety concern.

6 Case studies

The evolution and complexity of pig disease can be challenging. Knowledge of diseases continues to change as production practices and pathogens change, requiring veterinarians, nutritionists and animal scientists to be constantly attentive while monitoring the health and well-being of pigs.

PCV2v serves as a perfect case study to demonstrate these points and the complexity of evolution in knowledge and the pathogen itself. PCV2v was first identified in the late 1990s. This new pathogen appeared to be causing post-weaning wasting in several European countries while in the United States, most pigs tested positive for antibodies to the disease but otherwise were unaffected. The development and availability of new diagnostic techniques (antibody detection via ELISA) left veterinarians and pig farmers unsure of how to interpret the results, some even publically mocked the new discovery calling it a 'Circus' virus. Then in 2007–8 there was a genotype shift from PCV2a to PCV2b, which resulted in a highly pathogenic strain causing post-weaning wasting in pigs in North America (Carman et al., 2008). This small mutation had now changed the virus from a 'routine commensal' to a devastating, highly virulent, systemic wasting disease with herd mortalities at times in excess of 50%. In the same pen one would find pigs starting to waste right next to healthy-looking pigs (Fig. 2). All the pigs were eating, yet those infected pigs would quickly waste until they were euthanized or they would die. There was no halting the process. Blood

Figure 2 Size variation appearing in a group of field pigs infected with the new PCV2b variant starting to cause wasting in some of these pigs while others appear to be perfectly normal.

samples would show that most pigs (healthy and wasting ones) had antibodies and virus in their blood. It was no longer a matter of just knowing they were positive or negative. A new definition had to be developed to help clarify which pigs had PMWS and which did not. A diagnosis of PMWS now required three components:

1 lymphoid depletion,

2 large number of PCV2v in the lesion, and

3 clinical signs of wasting with a doubling in mortality (Sorden, 2000).

Then the miracle of vaccination came. Although pigs were being affected by PCV2b, the new vaccine used a PCV2a strain. Initially pig farmers and veterinarians were unsure about vaccinating their pigs against the old 'less pathogenic' strain when it was the new variant causing the high mortalities. The structure of the pig industry in the United States facilitated the spread of the PCV2b strain over the entire continental United States within just a few months. There was a new disease with new technology (diagnostics and vaccine), new knowledge and a new industry structure. The disease triad discussed in the introduction of this chapter (Fig. 1) was in full effect. Fortunately, it was quickly realized the new PCV2 vaccine truly was a miracle. The killed bacterin reduced mortalities from more than 25 to between 4 and 6% instantly; vaccinated pigs were now protected. Academically, it appeared the vaccine efficacy sounded too good to be true, but it was. Even herds that did not have the clinical disease and had 'normal' productivity who started using the PCV2 vaccine noted slight improvements in the overall health of the herd and lower mortalities.

7 Summary

Hopefully through reading this chapter the complexity of the various diseases has been emphasized. Diseases do not occur in a vacuum. They are impacted directly by a large cohort of factors including environment, nutrition, animal husbandry, genetics and co-infections amongst others.

Ultimately scientists, veterinarians, pig farmers, nutritionists, researchers and pig lovers are all striving to improve the health and well-being of the pigs raised so as to provide a more sustainable, abundant, wholesome, safe, economical and delicious protein for mankind.

8 Future trends

The future of sustainable pig production is positive. Vast amounts of knowledge are being gained rapidly. New technologies in diagnostic surveillance such as spatial-temporal pen sampling with oral fluids, metagenomics and microbiota in pig health are already being developed. Caution must be exercised when applying these new technologies to ensure a better understanding of what is known as well as what is unknown. A point of information overload is being reached as well as at times over-interpretation of the information. To ensure farmers stay focused on the ultimate goal, it is critical to collaborate with others who have expertise in different fields while embracing the clinical significance and implementation of new discoveries.

9 Where to look for further information

There are several sources available for additional information on swine diseases. *Diseases of Swine*, which is currently in its 10th edition, is recognized as the most comprehensive and authoritative textbook on swine diseases. Additionally the following websites can be consulted for information: The Pig Site (http://www.thepigsite.com/diseaseinfo/), Pig333 (https://www.pig333.com/pig-diseases/), The Merck Veterinary Manual (http://www.merckvetmanual.com/) and American Association of Swine Veterinarians Swine Disease Manual (https://vetmed.iastate.edu/vdpam/about/food-supply/swine/swine-disease-manual).

10 References

Alexandersen, S., Knowles, N. J., Dekker, A., Belsham, G. J., Zhang, Z. and Koenen, F. (2012) Picornaviruses. In: Zimmerman, J. J., Karriker, L. A., Ramirez, A., Schwartz, K. J. and Stevenson, G. W. (eds), *Diseases of* Swine, 10th ed. Ames: Wiley-Blackwell Publishing, pp 592.

Amtsberg, G. (1979) Determination of exfoliation triggering substances in cultures of *Staphylococcus hyicus* in swine and *Staphylococcus epidermidis* biotype 2 in cattle. *Zentralbl Veterinarmed B*. 26(4):257–72. German.

Arasteh, K. N., Cordes, C., Ewers, M., Simon, V., Dietz, E., Futh, U. M., Brockmeyer, N. H. and L'Age, P. (2000) Human immunodeficiency virus-related nontuberculous mycobacterial infection: incidence, survival analysis and associated risk factors. *Eur. J. Med. Res.* 5:424–30.

Armand-Lefevre, L., Ruimy, R. and Andremont, A. (2005) Clonal comparison of *Staphylococcus aureus* isolates from healthy pig farmers, human controls, and pigs. *Emerg. Infect. Dis.* 11(5):711–4.

Arruda, P. H., Madson, D. M., Ramirez, A., Rowe, E., Lizer, J. T. and Songer, J. G. (2013) Effect of age, dose and antibiotic therapy on the development of *Clostridium difficile* infection in neonatal piglets. *Anaerobe.* 22:104–10.

Bachrach, H. L. (1968) Foot-and-mouth disease. *Ann. Rev. Microbiol.* 22:201–44.

Bak, H. and Rathkjen, P. H. (2009) Reduced use of antimicrobials after vaccination of pigs against porcine proliferative enteropathy in a Danish SPF herd. *Acta Vet. Scand.* 51:1.

Barnes, D. M. and Sorensen, D. K. (1975) Salmonellosis. In: Dunne, H. W. and Leman, A. D. (eds), *Diseases of Swine*, 4th ed. Ames: Iowa State University Press, pp. 560–61.

Barrett, J. (1976) Studies on the induction of permeability in *Ascaris lumbricoides* eggs. *Parasitology* 73:109–21.

Bartlett, J. G., Chang, T. W., Gurwith, M., Gorbach, S. L. and Onderdonk, A. B. (1978) Antibiotic-associated pseudomembranous colitis due to toxin-producing clostridia. *N. Engl. J. Med.* 298(10):531–4.

Basi, C. (2015) Pigs that are resistant to incurable disease developed at University of Missouri. *New Bureau – University of Missouri.* [Online] 8 December. Available from: http://munews.missouri.edu/news-releases/2015/1208-pigs-that-are-resistant-to-incurable-disease-developed-at-university-of-missouri/ [Accessed 12 December 2015].

Baums, C. G. and Valentin-Weigand, P. (2009) Surface-associated and secreted factors of *Streptococcus suis* in epidemiology, pathogenesis and vaccine development. *Anim. Health Res. Rev.* 10(1):65–83.

Beier, D. and Gross, R. (2008) The BvgS/BvgA phosphorelay system of pathogenic *Bordetellae*: structure, function and evolution. *Adv. Exp. Med. Biol.* 631:149–60.

Bouchet, B., Vanier, G., Jacques, M. and Gottschalk, M. (2008) Interactions of *Haemophilus parasuis* and its LOS with Porcine Brain Microvascular Endothelial Cells. *Vet. Res.* 39:42.

Brochmeier, S. L., Halbur, P. G. and Thacker, E. L. (2002) Porcine Respiratory Disease Complex. In: Brogden, K. A. and Guthmiller, J. M. (eds). *Polymicrobial Diseases.* ASM Press:Washington DC.

Brochmeier, S. L., Register, K. B., Nicholson, T. L. and Loving, C. L. (2012) Bordetellosis. In: Zimmerman, J. J., Karriker, L. A., Ramirez, A., Schwartz, K. J. and Stevenson, G. W. (eds), *Diseases of Swine*, 10th ed. Ames: Wiley-Blackwell Publishing.

Brockmeier, S. L., Palmer, M. V., Bolin, S. R. and Rimler, R. B. (2001) Effects of intranasal inoculation with *Bordetella bronchiseptica*, porcine reproductive and respiratory syndrome virus, or a combination of both organisms on subsequent infection with *Pasteurella multocida* in pigs. *Am. J. Vet. Res.* 62(4):521–5.

Bueschel, D. M., Jost, B. H., Billington, S. J., Trinh, H. T. and Songer, J. G. (2003) Prevalence of cpb2, encoding beta2 toxin, in *Clostridium perfringens* field isolates: correlation of genotype with phenotype. *Vet. Microbiol.* 94:121.

Buller, C. R. and Moxley, R. A. (1988) Natural infection of porcine ileal dome M cells with rotavirus and enteric adenovirus. *Vet. Pathol.* 25:516–17.

Burrough, E. R., Strait, E. L., Kinyon, J. M., Bower, L. P., Madson, D. M., Wilberts, B. L., Schwartz, K. J., Frana, T. S. and Songer, J. G. Comparative virulence of clinical *Brachyspira spp.* isolates in inoculated pigs. *J. Vet. Diagn. Invest.* 24(6):1025-34.

Calsamiglia, M. and Pijoan, C. (2000) Colonisation state and colostral immunity to *Mycoplasma hyopneumoniae* of different parity sows. *Vet. Rec.* 146:530-2.

Carlson, S. A., Barnhill, A. E. and Griffith, R. W. (2012) Salmonellosis. In: Zimmerman, J. J., Karriker, L. A., Ramirez, A., Schwartz, K. J. and Stevenson, G. W. (eds), *Diseases of Swine*, 10th ed. Ames: Wiley-Blackwell Publishing.

Carman, S., Cai, H. Y., DeLay, J., Youssef, S. A., McEwen, B. J., Gagnon, C. A., Tremblay, D., Hazlett, M., Lusis, P., Fairles, J., Alexander, H. S. and van Dreumel, T. (2008) The emergence of a new strain of porcine circovirus-2 in Ontario and Quebec swine and its association with severe porcine circovirus associated disease-2004-2006. *Can. J. Vet. Res.* 72:259-68.

Carter, G. R. (1955) Studies on *Pasteurella multocida*: A hemagglutination test for the identification of serological types. *Am. J. Vet. Res.* 16:481-4.

CFSPH - Center for Food Security and Public Health. (2005) Tanea infections. [Online] Available from: http://www.cfsph.iastate.edu/Factsheets/pdfs/taenia.pdf. [Accessed 31 May 2016].

CFSPH - Center for Food Security and Public Health. (2014) *Foot and mouth disease*. [Online] Available from: http://www.cfsph.iastate.edu/Factsheets/pdfs/foot_and_mouth_disease.pdf. [Accessed: 17th December 2015].

CFSPH - Center for Food Security and Public Health. (2015a) *African Swine Fever*. [Online] Available from: http://www.cfsph.iastate.edu/Factsheets/pdfs/african_swine_fever.pdf. [Accessed: 10th January 2016].

CFSPH - Center for Food Security and Public Health. (2015b) *Classical Swine Fever*. [Online] Available from: http://www.cfsph.iastate.edu/Factsheets/pdfs/classical_swine_fever.pdf. [Accessed: 22nd January 2016].

Chander, Y., Primus, A., Oliveira, S. and Gebhart, C. J. (2012) Phenotypic and molecular characterization of a novel strongly hemolytic *Brachyspira* species, provisionally designated '*Brachyspira hampsonii*'. *J. Vet. Diagn. Invest.* 24(5):903-10.

Chang, K.-O., Saif, L. J. and Kim, Y. (2012) Reoviruses (Rotaviruses and Reoviruses). In: Zimmerman, J. J., Karriker, L. A., Ramirez, A., Schwartz, K. J. and Stevenson, G. W. (eds), *Diseases of Swine*, 10th ed. Ames: Wiley-Blackwell Publishing.

Chianini, F., Majo, N., Segales, J., Dominguez, J. and Domingo, M. (2003) Immunohistochemical characterization of PCV2 associate lesions in lymphoid and non-lymphoid tissues of pigs with natural postweaning multisystemic wasting syndrome (PMWS). *Vet. Immunol. Immunopathol.* 94:63-75.

Chiers, K., De Waele, T., Pasmans, F., Ducatelle, R. and Haesebrouck, F. (2010) Virulence factors of *Actinobacillus pleuropneumoniae* involved in colonization, persistence and induction of lesions in its porcine host. *Vet. Res.* 41(5):65.

Christensen, L. S., Mousing, J., Mortensen, S., Soerensen, K. J., Strandbygaard, S. B., Henriksen, C. A. and Andersen, J. B. (1990) Evidence of long distance airborne transmission of Aujeszky's disease (pseudorabies) virus. *Vet. Rec.* 127:471-4.

Clota, J., Foix, A., March, R., Riera, P. and Costa, L. (1996) Caracterización serológica de cepas de *Actinobacillus pleuropneumoniae* aisladas en Espaa. *Med. Vet.* 13:17-22.

Colgrove, G. S., Haelterman, E. O. and Coggins, L. (1969) Pathogenesis of African swine fever in young pigs. *Am. J. Vet. Res.* 30(8):1343-59.

Cottral, G. E. (1969) Persistence of foot-and-mouth disease virus in animals, their products and the environment. *Bull.Off. Int. Epizoot.* 70:549–68.

De Boer, C. V. (1967) Studies to determine neutralizing antibody in sera from animals recovered from African swine fever and laboratory animals inoculated with African swine fever virus with adjuvants. *Arch Gesamte Virusforsch.* 20:164–79.

de Jong, M. F. and Nielsen, J. P. (1990) Definition of progressive atrophic rhinitis. *Vet. Rec.* 126:93.

De Villier, E. P., Gallardo, C., Arias, M., Da Silva, M., Upton, C., Martin, R. and Bishop, R. P. (2010) Phylogenomic analysis of 11 complete African swine fever virus genome sequences. *Virol.* 400(1):128–36.

Dee, S. A., Joo, H. S., Henry, S., Tokach, L., Park, K., Molitor, T. and Pijoan, C. (1996) Detecting subpopulations after PRRS virus infection in large breeding herds using multiple serologic tests. *J. Swine Health Prod.* 4(4):181–4.

Depner, K. R., Hinrichs, U., Bickhardt, K., Greiser-Wilke, I., Pohlenz, J., Moennig, V. and Liess, B. (1997) Influence of breed-related factors on the course of classical swine fever virus infection. *Vet. Rec.* 140:506–7.

Desrosiers, R. (2004) Epidemiology, diagnosis and control of swine diseases. Howard Dunne Memorial Lecture. In *Proc. 35th Annu. Meet Am. Assoc. Swine Pract.*, pp. 9–37.

Domenech, J., Lubroth, J. and Sumption, K. (2010) Immune protection in animals: the examples of rinderpest and foot-and-mouth disease. *J. Comp. Pathol.* 142 Suppl 1:S120–4.

Doyle, L. (1932) A rickettsia-like or anaplasmos-like disease in swine. *J. Am. Vet. Med. Assoc.* 8:668–71.

Duan, X., Nauwynck, H. J. and Pensaert, M. B. (1997) Virus quantification and identification of cellular targets in the lungs and lymphoid tissues of pigs at different time intervals after inoculation with porcine reproductive and respiratory syndrome virus (PRRSV). *Vet. Microbiol.* 56(1–2):9–19.

European Union Food Safety Authority. (2011) Scientific Opinion on the public health hazards to be covered by inspection of meat (swine). *EFSA Journal.* 9(10):2351.

Fairbrother, J. M. and Nadeau, E. (2006) *Escherichia coli*: on-farm contamination of animals. *Rev. Sci. Tech.* 25(2):555–69.

Farez, S. and Morley, R. S. (1997) Potential animal health hazards of pork and pork products. *Rev. Sci. Tech.* 16(1):65–78.

Farzan, A., Kircanski, J., DeLay, J., Soltes, G., Songer, J. G., Friendship, R. and Prescott, J. F. (2013) An investigation into the association between cpb2-encoding *Clostridium perfringens* type A and diarrhea in neonatal piglets. *Can. J. Vet. Res.* 77(1):45–53.

Foley, S. L., Lynne, A. M. and Nayak, R. Salmonella challenges: prevalence in swine and poultry and potential pathogenicity of such isolates. *J. Anim. Sci.* 2008 April; 86(14 Suppl):E149–62. Epub 2 October 2007. Review. PubMed PMID: 17911227.

Fort, M., Olvera, A., Sibila, M., Segalés, J. and Mateu, E. (2007) Detection of neutralizing antibodies in postweaning multisystemic wasting syndrome (PMWS)-affected and non-PMWS-affected pigs. *Vet. Microbiol.* 125:244–55.

Frana, T. S. (2012) Staphylococcosis. In: Zimmerman, J. J., Karriker, L. A., Ramirez, A., Schwartz, K. J. and Stevenson, G. W. (eds) *Diseases of Swine*, 10th ed. Ames: Wiley-Blackwell Publishing.

Frana, T. S., Beahm, A. R., Hanson, B. M., Kinyon, J. M., Layman, L. L., Karriker, L. A., Ramirez, A. and Smith, T. C. (2013) Isolation and characterization of methicillin-resistant

Staphylococcus aureus from pork farms and visiting veterinary students. *PLoS One.* 8(1):e53738.

Frey, J., Haldimann, A. and Nicolet, J. (1992) Chromosomal heterogeneity of various *Mycoplasma hyopneumoniae* field strains. *Int. J. Syst. Bacteriol.* 42:275-80.

Friis, N. F., Hansen, K. K., Schirmer, A. L. and Aabo, S. (1992) *Mycoplasma hyosynoviae* in joints with arthritis in abattoir baconers. *Acta Vet. Scand.* 33:425-9.

Goldberg, T. L., Lowe, J. F., Milburn, S. M. and Firkins, L. D. (2003) Quasispecies variation of porcine reproductive and respiratory syndrome virus during natural infection. *Virology.* 20; 317(2):197-207.

Goodwin, R. F. (1985) Apparent reinfection of enzootic-pneumonia-free pig herds: search for possible causes. *Vet. Rec.* 116:690-4.

Gottschalk, M. (2012) Actinobacillosis. In: Zimmerman, J. J., Karriker, L. A., Ramirez, A., Schwartz, K. J. and Stevenson, G. W. (eds), *Diseases of Swine*, 10th ed. Ames: Wiley-Blackwell Publishing.

Gottschalk, M. (2012) Streptococcosis. In: Zimmerman, J. J., Karriker, L. A., Ramirez, A., Schwartz, K. J. and Stevenson, G. W. (eds), *Diseases of Swine*, 10th ed. Ames: Wiley-Blackwell Publishing.

Gottschalk, M., Broes, A. and Fittipaldi, N. (2003) Recent developments on *Actinobacillus pleuropneumoniae*. In *Proc. 34th Annu. Meet Am. Assoc. Swine Pract*, pp. 387-93.

Greve, J. H. (2012) Internal parasites: Helminths. In: Zimmerman, J. J., Karriker, L. A., Ramirez, A., Schwartz, K. J. and Stevenson, G. W. (eds), *Diseases of Swine*, 10th ed. Ames: Wiley-Blackwell Publishing.

Groebel, K., Hoelzle, K., Wittenbrink, M. M., Ziegler, U. and Hoelzle, L. E. (2009) *Mycoplasma suis* invades porcine erythrocytes. *Infect. Immun.* 77:576-84.

Hammer, J. M. (2004) The temporal relationship of fecal shedding of *Lawsonia intracellularis* and seroconversion in field cases. *J. Swine Health Prod.* 12: 29-33.

Hampson, D. (2012) Brachyspiral Colitis. In: Zimmerman, J. J., Karriker, L. A., Ramirez, A., Schwartz, K. J. and Stevenson, G. W. (eds), *Diseases of Swine*, 10th ed. Ames: Wiley-Blackwell Publishing.

Harding, J. C. and Clark, E. G. (1997) Recognizing and diagnosing postweaning multisystemic wasting syndrome (PMWS). *J. Swine Health Prod.* 5:201-3.

Hill, J. E., Lomax, L. G., Lindsay, D. S. and Lynn, B. S. (1985) Coccidosis caused by *Eimeria scabra* in a finishing hog. *J. Am. Vet. Med. Assoc.* 186(9):981-3.

Jung, K., Hu, H. and Saif, L. J. (2016) Porcine deltacoronavirus induces apoptosis in swine testicular and LLC porcine kidney cell lines in vitro but not in infected intestinal enterocytes in vivo. *Vet. Microbiol.* 182:57-63.

Jung, K. and Saif, L. J. (2015) Porcine epidemic diarrhea virus infection: Etiology, epidemiology, pathogenesis and immunoprophylaxis. *Vet. J.* 204(2):134-43.

Kelly, A. R., Jones, R. J., Gillick, J. C. and Sims, L. D. (1984) Outbreak of botulism in horses. *Eq. Vet. J.* 16:519-21.

Kessler, E., Matthes, H. F., Schein, E. and Wendt, M. (2003) Detection of antibodies in sera of weaned pigs after contact infection with Sarcoptes scabiei var. suis and after treatment with an antiparasitic agent by three different indirect ELISAs. *Vet. Parasitol.* 114(1):63-73.

Kielstein, P. and Rapp-Gabrielson, V. J. (1992) Designation of 15 serovars of *Haemophilus parasuis* on the basis of immunodiffusion using heat-stable antigen extracts. *J. Clin. Microbiol.* 30(4):862-5.

Kitching, R. P., Knowles, N. J., Samuel, A. R. and Donaldson, A. I. (1989) Development of foot-and-mouth disease virus strain characterisation--a review. *Trop. Anim. Health Prod.* 21:153–66.

Kucerova, Z., Jaglic, Z., Ondriasova, R. and Nedbalcova, K. (2005) Serotype distribution of *Actinobacillus pleuropneumoniae* isolated from porcine pleuropneumonia in the Czech Republic during period 2003-2004. *Vet. Med. – Czech.* 50:355-60.

Leiser, O. P., Corn, J. L., Schmit, B. S., Keim, P. S. and Foster, J. T. (2013) Feral swine brucellosis in the United States and prospective genomic techniques for disease epidemiology. *Vet. Microbiol.* 166(1–2):1–10.

Leuenberger, R., Boujon, P., Thür, B., Miserez, R., Garin-Bastuji, B., Rüfenacht, J. and Stärk, K. D. (2007) Prevalence of classical swine fever, Aujeszky's disease and brucellosis in a population of wild boar in Switzerland. *Vet. Rec.* 160(11):362–8.

Li, G., Chen, Q., Harmon, K. M., Yoon, K. J., Schwartz, K. J. and Hoogland, M. J. (2014) Full-length genome sequence of porcine deltacoronavirus strain USA/IA/2014/8734. *Genome Announc.* 2:e00278-14.

Lindsay, D. S., Blagburn, B. L. and Boosinger, T. R. (1987) Experimental *Eimeria debliecki* infections in nursing and weaned pigs. *Vet. Parasitol.* 25(1):39–45.

Magstadt, D., Madson, D., Arruda, P., Hoag, H., Sun, D., Pillatzki, A., Stevenson, G., Wilberts, B., Brodie, J., Harmon, K., Chong, W. and Main, R. (2014) Porcine epidemic diarrhea: Pathogenesis and viremia in weaned and neonatal pigs. *In Proc 22nd Annual Swine Disease Conference for Swine Practitioners*, pp. 34-7.McOrist, S. (2005) Defining the full cost of endemic porcine proliferative enteropathy. *Vet. J.* 170(1):8–9.

McOrist, S., Smith, S. H. and Green, L. J. (1997) Estimate of direct financial losses due to porcine proliferative enteropathy. *Vet. Rec.* 140: 579–81.

McOrist, S., Smith, S. H. and Klein, T. (1999) Monitored control programme for proliferative enteropathy on British pig farms. *Vet. Rec.* 144: 202–4.

Mebus, C. A., House, C., Ruiz Gonzalvo, F., Pineda, J. M., Tapiador, J., Pire, J. J., Bergada J., Yedloutschnig, R. J., Sahu, S., Becerra, V. and Sanchez-Vizcaino, J. M. (1993) Survival of foot-and-mouth disease, African swine fever, and hog cholera viruses in Spanish serrano cured hams and Iberian cured hams, shoulders and loins. *Food Micro.* 10:133-43.

Médici, K. C., Barry, A. F., Alfieri, A. F. and Alfieri, A. A. (2011) Porcine rotavirus groups A, B, and C identified by polymerase chain reaction in a fecal sample collection with inconclusive results by polyacrylamide gel electrophoresis. *J. Swine Health Prod.* 19(3):146-50.

Mittal, K. R., Higgins, R., Larivière, S. and Nadeau M. (1992) Serological characterization of *Actinobacillus pleuropneumoniae* strains isolated from pigs in Quebec. *Vet. Microbiol.* 32(2):135-48.

Moennig, V., Floegel-Niesmann, G. and Greiser-Wilke, I. (2003) Clinical signs and epidemiology of classical swine fever: A review of new knowledge. *Vet. J.* 165:11-20.

Nelson, M. I. and Vincent, A. L. (2014) Reverse zoonosis of influenza to swine: new perspectives on the human-animal interface. *Trends Microbiol.* 23(3):142–53.

Oh, Y., Han, K., Seo, H. W., Park, C. and Chae, C. (2013) Program of vaccination and antibiotic treatment to control polyserositis caused by Haemophilus parasuis under field conditions. *Can. J. Vet. Res.* 77(3): 183–90.

Oliveira, S., Pijoan, C. and Morrison, R. (2004) Evaluation of *Haemophilus parasuis* control in the nursery using vaccination and controlled exposure. *J. Swine Health Prod.* 12(3):123-8.

Olsen, S. C., Garin-Bastuji, B., Blasco, J. M., Nicola, A. M, and Samartino, L. (2012) Brucellosis In: Zimmerman, J. J., Karriker, L. A., Ramirez, A., Schwartz, K. J. and Stevenson, G. W. (eds), *Diseases of Swine*, 10th ed. Ames: Wiley-Blackwell Publishing.

Otake, S., Dee, S., Corzo, C., Oliveira, S. and Deen, J. Long-distance airborne transport of infectious PRRSV and *Mycoplasma hyopneumoniae* from a swine population infected with multiple viral variants. *Vet. Microbiol.* 145(3-4):198-208.

Pappas, G., Akritidis, N., Bosilkoviski, M. and Tsianos, E. (2005) Brucellosis. *N. Engl. J. Med.* 352:2325-36.

Pijoan, C., Morrison, R. B. and Hilley, H. D. (1983) Serotyping of *Pasteurella multocida* isolated from swine lungs collected at slaughter. *J. Clin. Microbiol.* 17:1074-6.

Piriou, L., Chevallier, S., Hutet, E., Charley, B., Le Potier, M. F. and Albina, E. (2003) Humoral and cell-mediated immune responses of d/d histocompatible pigs against classical swine fever (CSF) virus. *Vet. Res.* 34:389-404.

Ploug, T., Holm, S. and Gjerris, M. (2015) The stigmatization dilemma in public health policy-the case of MRSA in Denmark. *BMC Public Health.* 15:640.

Pogranichny, R. M., Yoon, K. J., Harms, P. A., Swenson, S. L., Zimmerman, J. J. and Sorden, S. D. (2000) Characterization of immune response of young pigs to porcine circovirus type 2 infection. *Viral. Immunol.* 13:143-53.

Ramirez, A., Rowe, E. W., Arruda, P. H. and Madson, D. (2014) Use of equine-origin antitoxins in piglets prior to exposure to mitigate the effects of *Clostridium difficile* infection – a pilot study. *J. Swine Health Prod.* 2014, 22(1):29-32.

Rimler, R. B. and Rhoades, K. R. (1987) Serogroup F, a new capsule serogroup of *Pasteurella multocida. J. Clin. Microbiol.* 25:615-18.

Ross, R. F. and Whittlestone, P. (1983) Recovery of, identification of, and serological response to porcine mycoplasmas. In: Tully, J. G. and Razin, S. (eds), *Methods in Mycoplasmology*. New York: Academic Press.

Rowland, R. R., Lunney, J. and Dekkers, J. (2012) Control of porcine reproductive and respiratory syndrome (PRRS) through genetic improvements in disease resistance and tolerance. *Front. Genet.* 3:260.

Rowland, R. R., Steffen, M., Ackerman, T. and Benfield, D. A. (1999) The evolution of porcine reproductive and respiratory syndrome virus: quasispecies and emergence of a virus subpopulation during infection of pigs with VR-2332. *Virology.* 5, 259(2):262-6.

Ruiz Gonzalvo, F., Carnero, M. E. and Bruyel, V. (1981) Immunological responses of pigs to partially attenuated ASF and their resistance to virulent homologous and heterologous viruses. In: Wilkinson, P. J. (ed.), *FAO/CEC Expert Consultation in ASF Research*. Rome, pp. 206-16.

Saif, L. J. (1999) Comparative pathogenesis of enteric viral infections of swine. *Adv. Exp. Med. Biol.* 473:47-59.

Saif, L. J., van Cott, J. L. and Brim, T. A. (1994) Immunity to transmissible gastroenteritis virus and porcine respiratory coronavirus infections in swine. *Vet. Immunol. Immunopathol.* 43(1-3):89-97.

Sanford, S. E. and Tilker, M. E. (1982) *Streptococcus suis* type II-associated diseases in swine: observations of a one-year study. *J. Am. Vet. Med. Assoc.* 181(7):673-6.

Schalk, A. F., Roderick, L. M., Foust, H. L. and Harshfield, G. S. (1935) Avian Tuberculosis: Collected Studies. *ND Agric. Exp. Stn. Tech. Bull.* 279.

Schelkopf, A. C., Magstadt, D. R., Arruda, B. L. Arruda, P. H. E., Zimmerman, J. J. Wetzell, T., Dee, S. A. and Madson, D. M. (2015) Lactogenic Protection against porcine epidemic diarrhea virus in piglets following homologous challenge. *In Proc 23rd Annual Swine Disease Conference for Swine Practitioners*, pp. 41-4.

Schliesser, T. and Weber, A. (1973) Untersuchungen über die Tenazitzt von Mykobakterien der Gruppe III nach Runyon in Sagemehleinstreu. *Zentralbl Veterinärmed (B)* 20:710-14.

Schwartz, K. J. 1997. Salmonellosis. Proc AASP, Quebec City, Quebec.

Sjölund, M., de la Fuente, A., Fossum, C. and Wallgren, P. (2009) Responses of pigs to a re-challenge with *Actinobacillus pleuropneumoniae* after being treated with different antimicrobials following their initial exposure. *Vet Rec.* 164:550-5.

Smith, H. J. (1986) Transmission of *Sarcoptes scabiei* in Swine by Fomites. *Can. Vet J.* 27(6):252-54.

Sorden, S. D. (2000) Update on porcine circovirus and postweaning multisystemic wasting syndrome. *J. Swine Health Prod.* 8:133-6.

Stuart, B. P. and Lindsay, D. S. (1986) Coccidiosis in swine. *Vet. Clin. North Am. Food Anim. Pract.* 2(2):455-68.

Thacker, E. J. and Minion, F. C. (2012) Mycoplasmosis. In: Zimmerman, J. J., Karriker, L. A., Ramirez, A., Schwartz, K. J. and Stevenson, G. W. (eds), *Diseases of Swine*, 10th ed. Ames: Wiley-Blackwell Publishing.

Thanawongnuwech, R., Thacker, E. L. and Halbur, P. G. (1997) Effect of porcine reproductive and respiratory syndrome virus (PRRSV) (isolate ATCC VR-2385) infection on bactericidal activity of porcine pulmonary intravascular macrophages (PIMs): in vitro comparisons with pulmonary alveolar macrophages (PAMs). *Vet. Immunol. Immunopathol.* 59(3-4):323-35.

Theil, K. W., Bohl, E. H., Cross, R. F., Kohler, E. M. and Agnes, A. G. (1978) Pathogenesis of porcine rotaviral infection in experimentally inoculated gnotobiotic pigs. *Am. J. Vet. Res.* 39:213-20.

Thoen, C. O., Jarnagin, J. L. and Richards, W. D. (1975) Isolation and identification of mycobacteria from porcine tissues: A three-year summary. *Am. J. Vet. Res.* 36:1383-6.

Thoen, C. O. (2012) Tuberculosis. In: Zimmerman, J. J., Karriker, L. A., Ramirez, A., Schwartz, K. J. and Stevenson, G. W. (eds), *Diseases of Swine*, 10th ed. Ames: Wiley-Blackwell Publishing.

Thomas, P. (2014) Field experiences using porcine epidemic diarrhea virus (PEDv) vaccine in herds experiencing endemic disease. *In Proc 22nd Annual Swine Disease Conference for Swine Practitioners*, pp. 38-42.

Torremorrell, M., Moore, C. and Christianson, W. T. (2002) Establishment of a herd negative for porcine reproductive and respiratory syndrome virus (PRRSV) from PRRSV-positive sources. *J. Swine Health Prod.* 10(4):153-60.

United States Department of Agriculture. (2011) *News Release*. [Online] Available from: http://www.fsis.usda.gov/wps/portal/fsis/newsroom/!ut/p/a0/04_ Sj9CPykssy0xPLMnMz0vMAfGjzOINAg3MDC2dDbwMDIHQ08842MTDy8_YwN tMvyDbUREAzbjixQ!!/?1dmy&current=true&urile=wcm%3Apath%3A%2FF SIS-Archives-Content%2Finternet%2Fmain%2Fnewsroom%2Fnews-releases-statements-and-transcripts%2Fnews-release-archives-by-year%2Farchives%2FCT_ Index919. [Accessed: 31st May 2016].

Vanier, G., Szczotka, A., Friedl, P., Lacouture, S., Jacques, M. and Gottschalk, M. (2006) *Haemophilus parasuis* invades porcine brain microvascular endothelial cells. *Microbiology.* 152:135-42.

Wagstrom, E. A., Yoon, K. J. and Zimmerman, J. J. (2000) Immune components in porcine mammary secretions. *Viral. Immunol.* 13:383-97.

Wang, L., Byrum, B. and Zhang, Y. (2014) Detection and genetic characterization of deltacoronavirus in pigs, Ohio, USA, 2014. *Emerg. Infect Dis.* 20:1227-30.

Waters, M., Savoie, A., Garmory, H. S., Bueschel, D., Popoff, M. R., Songer, J. G., Titball, R. W., McClane, B. A. and Sarker, M. R. (2003) Genotyping and phenotyping of beta2-toxigenic *Clostridium perfringens* fecal isolates associated with gastrointestinal diseases in piglets. *J. Clin. Microbiol.* 41:3584-91.

Wertheim, H. F., Nghia, H. D., Taylor, W. and Schultsz, C. (2009) *Streptococcus suis*: an emerging human pathogen. *Clin. Infect. Dis.* 48(5):617-25.

Wiseman, B., Harris, D. L., Glock, R. D. and Wilkins, L. (1989) Management of seedstock that is negative for *Haemophilus parasuis*. In *Proc 20th Annu Meet Am Assoc Swine Pract.*, pp. 23-5.

Wittmann, G. and Rziha, H. J. (1989) Aujeszky's disease (pseudorabies) in pigs. In: Wittmann, G. (ed.), *Herpesvirus diseases of cattle, horses and pigs*. Boston: Kluwer Academic Publishers.

World Health Organization. (2016) *Taeniasis/cysticercosis*. [Online] Available from: http://www.who.int/mediacentre/factsheets/fs376/en/. [Accessed: 26th May 2016].

World Organization for Animal Health and Food and Agriculture Organization of the United Nations. (2012) The Global Foot and Mouth Disease Control Strategy - Strengthening animal health systems through improved control of major disease. [Online] Available from: http://www.oie.int/doc/ged/D11886.PDF. [Accessed: 26th May 2016].

Yaeger, M., Funk, N. and Hoffman, L. (2002) A survey of agents associated with neonatal diarrhea in Iowa swine including *Clostridium difficile* and porcine reproductive and respiratory syndrome virus. *J. Vet. Diagn. Invest.* 14(4):281-7.

Zhang, J., Chen, Q., Thomas, J. and Gauger, P. (2015) Characterization of pathogenicity and cross-protective immunity of U. S. PEDVs. *In Proc 23rd Annual Swine Disease Conference for Swine Practitioners*, pp. 45-9.

Zhang, Q., Young, T. F. and Ross, R. F. (1994) Microtiter plate adherence assay and receptor analogs for *Mycoplasma hyopneumoniae*. *Infect. Immun.* 62:1616-22.

Chapter 2

Understanding and identifying bacterial disease in swine

Dominiek Maes, Filip Boyen, and Freddy Haesebrouck, Ghent University, Belgium

1 Introduction

Advances in the development of diagnostic tools during the last decades such as polymerase chain reaction (PCR) have allowed to diagnose different viral infections in pigs faster and with a higher sensitivity compared to previous cumbersome techniques such as virus isolation. This increased detection rate of viruses might have contributed to the perception that bacteria are considered as pathogens that are only or mainly important following previous viral infection. However, infections with bacterial pathogens have a significant impact on pork production worldwide and may lead to clinical disease even in the absence of prior viral infections (Ramirez, 2018). They continue to significantly affect the health, well-being, and performance of pigs in all types of production systems. Infections with bacterial respiratory and enteric pathogens are among the most common and economically significant diseases in swine. *Mycoplasma hyopneumoniae* is the primary pathogen that causes enzootic pneumonia in pigs and an important pathogen within the porcine respiratory disease complex (PRDC) (Rycroft, 2020). *Escherichia coli* (*E. coli*) is among the most important

http://dx.doi.org/10.19103/AS.2022.0103.02

causes of diarrhea in suckling and recently weaned pigs (Fairbrother and Nadeau, 2019). Some swine bacterial diseases are also on the Office International des Epizooties (OIE) list (2021) of notifiable animal diseases such as anthrax, infections with *Brucella suis*, and the *Mycobacterium tuberculosis* complex.

Infections with bacterial pathogens are also responsible for a major part of the antimicrobial use and, therefore, indirectly increase the risk for the development of antimicrobial resistance. Worldwide, pig farmers and veterinarians are urged to use less antimicrobials responsibly and judiciously. This implies that antimicrobials should not be used for prophylaxis and that antimicrobials that are critically important for human medicine should not be used anymore in animals or only under very strict conditions.

Finally, some bacterial pathogens in swine, like *Salmonella enterica*, are important zoonotic agents. In the European Union (EU), salmonellosis is the second most reported zoonosis in humans (after campylobacteriosis), and *Salmonella* is the most frequently reported causative agent isolated in foodborne outbreaks (EFSA, 2018). Pigs and pork are considered as the source of 28% of the human infections caused by *Salmonella* Typhimurium and of 50% of the human infections caused by *Salmonella* Typhimurium monophasic variants (EFSA, 2018). Also in other parts of the world like North America, *Campylobacter* and *Salmonella* cause the largest proportion of foodborne illnesses (Tack et al., 2020).

The present chapter provides an overview of important swine bacterial pathogens, the mechanisms of pathogenicity, transmission routes, clinical signs and lesions caused by the pathogens, and control measures.

2 Some phenotypic characteristics of swine bacterial pathogens

Bacteria can be classified according to some basic phenotypic characteristics such as (1) Gram stain reaction, (2) morphology, (3) oxygen requirements, and (4) ability to form endospores (Table 1) (Post, 2019).

Gram staining differentiates bacteria by the chemical and physical properties of their cell walls. Gram-positive cells have a thick layer of peptidoglycan in the cell wall that retains the primary stain, crystal violet. They appear as purple bacteria. Gram-negative cells have a thinner peptidoglycan layer that allows the crystal violet to wash out upon the addition of ethanol. They are stained pink or red by the counterstain, commonly safranin or fuchsine. Mycoplasmas are of the gram-positive lineage, but due to the lack of a cell wall, they do not retain the crystal violet stain. Treponemes and leptospires and also *Mycobacterium* species with their high lipid content in the cell walls cannot be properly stained by this method, but other methods such as Ziehl-Neelsen staining for *Mycobacterium* species are more appropriate.

Table 1 Classification of the principal bacterial pathogens of swine

Gram-staining	Oxygen requirements	Spore formation	Shape	Genus/genera
Gram-positive	Aerobic to microaerophilic	No	Bacilli	*Trueperella, Listeria, Mycobacterium, Rhodococcus*
	Aerobic or facultatively anaerobic	Yes	Bacilli	*Bacillus*
	Facultatively anaerobic	No	Cocci	*Enterococcus, Staphylococcus, Streptococcus*
		No	Bacilli	*Erysipelothrix*
	Anaerobic	Yes	Bacilli	*Clostridium*
		No	Bacilli	*Actinobaculum*
Gram-negative	Aerobic or facultatively anaerobic	No	Bacilli	*Actinobacillus, Bordetella, Brucella, Burkholderia, Escherichia, Glaesserella, Pasteurella, Salmonella, Yersinia*
	Microaerophilic or anaerobic	No	Curved to spiral-shaped bacilli	*Brachyspira, Leptospira, Treponema, Borrelia, Campylobacter, Helicobacter, Lawsonia*
Bacteria without cell wall and obligately intracellular bacteria				*Mycoplasma, Chlamydia*

Source: Adapted from Post (2019).

The cellular morphology of bacteria comprises cocci, bacilli, and spiral-shaped bacteria. Some cocci, such as *Staphylococcus* species, are perfectly rounded, while others are somewhat elongated such as many *Streptococcus* and *Enterococcus* species, or may even look like rods (coccobacilli). Staphylococci usually occur in irregular groups and less frequently in the classical 'grapes' and also in short chains and in pairs. Only some streptococci form clear chains. Many streptococci and all enterococci occur more often in groups, grapes, in duplo, as singles, or in short chains.

Bacilli may appear as regular rods (members of the genera *Escherichia*, *Salmonella*, and *Listeria*), others as coccobacillary (*Pasteurella*), pleomorphic (*Trueperella pyogenes* [*T. pyogenes*] and *Actinobacillus* species), curved (*Campylobacter* species and *Lawsonia intracellularis* [*L. intracellularis*]) or filamentous in shape (*Actinomyces* species and *Nocardia* species). Bacteria with a spiral shape include *Brachyspira, Treponema, Borrelia*, and *Leptospira* species, and several *Helicobacter* species. The size of the bacterium depends upon its growth phase and the type of medium used for cultivation. In general, most spirochetes, *Bacillus* species, and clostridia are regarded as large pseudomonads; Enterobacteriaceae (*Salmonella* species and *E. coli*) as medium-sized; *Brucella*, *Glaesserella*, and *Pasteurella* species as small; and mycoplasmas and chlamydias as very small.

Bacteria can also be classified by their ability to utilize or tolerate oxygen. Bacteria that require oxygen and can tolerate levels of oxygen that are present in the atmosphere (21%) are called 'obligate aerobes.' Facultatively anaerobic bacteria can multiply in both the presence and absence of oxygen. This group represents the majority of the swine bacterial pathogens. Microaerophilic bacteria such as the genera *Campylobacter* and *Helicobacter* require small amounts of oxygen but may be killed by normal atmospheric concentrations. Obligate anaerobes can only grow in the absence of oxygen and several of these bacteria are killed by even small amounts of oxygen. Many clostridia are obligate anaerobes. Some obligate anaerobic species, like *Brachyspira hyodysenteriae*, can be more tolerant to oxygen and can therefore survive longer in the environment than other strictly anaerobic bacteria.

Bacteria may also be classified based on their ability to produce spores. Spores are formed when vegetative cells are deprived of necessary nutrition or environmental conditions. Spores are extremely resistant to harsh environmental conditions and disinfectants. Two important genera of spore-forming bacteria in swine are *Bacillus* and *Clostridium*.

The phenotypic characteristics of the bacteria may be important for the epidemiology, diagnostics, treatment, and control measures. The outcome of the Gram staining, for instance, may sometimes help to determine the choice of antimicrobial for treatment. Knowledge of a bacterium's oxygen requirements may be important for the storage and transport conditions of diagnostic samples to the laboratory. Organisms that produce spores, such as *Clostridia*, may be difficult to eliminate from the barns because the spores are resistant to most cleaning and disinfectant agents.

3 Advanced methods to identify and classify swine bacterial pathogens

Matrix-assisted laser desorption/ionization time-of-flight mass spectrometry (MALDI-TOF MS) is a technique which has become very popular in veterinary diagnostics in the last few years (Randall et al., 2015). An easy, cheap, and fast preparation protocol following a culture step allows identifying most bacterial species with high accuracy. The technique results in a peak spectrum representing the most abundant bacterial proteins, which is compared to a commercial and/or in-house library (Pavlovic et al., 2013).

The mentioned phenotypic characteristics and the traditional methods for identifying microbes are still used but have some drawbacks. First, they are applicable mainly to organisms that can be cultured *in vitro*. As the growth requirements of various species of bacteria are unknown, methods based on cultivation are not sufficient to explore specific microbial populations like the microbiome in the gastrointestinal tract (Lamendella et al., 2011). Second, some bacterial strains might exhibit unique biochemical characteristics that do not fit the pattern of any known genus and species. With the advent of molecular phylogeny and taxonomy, more advanced methods are used to identify and classify bacterial pathogens. They are not dependent on live cultures, and they can often reveal minute differences between organisms that escape detection by traditional means.

Several genes are used to identify species, the most important being the *16S* rRNA gene, followed by *23S*, internal transcribed spacer (ITS) region, gyrase B (*gyrB*), and other, sometimes species-specific, genes to attain a better resolution between species (Klindworth et al., 2013). In addition to MALDI-TOF MS, one of the quickest ways to match an isolated strain to a species or genus is done by amplifying its *16S* gene with universal primers, sequencing the 1.4 kb amplicon, and submitting it to a specialized web-based identification database. Several identification methods exist such as phenotypic analyses (fatty acid analyses, growth conditions), genetic analyses (DNA–DNA hybridization, DNA profiling, sequence, GC ratios), and phylogenetic analyses (*16S*-based phylogeny, phylogeny based on other genes, multi-gene sequence analysis, and whole-genome sequence-based analysis). Whole-genome sequencing is particularly useful for characterizing isolates, predicting phenotypes, tracking outbreaks, and identifying infection sources (Balloux et al., 2018).

4 Pathogenicity of swine bacteria

Most bacteria associated with pigs do not cause disease. These nonpathogenic, commensal, or mutualistic bacteria belong to the normal microbiota present on the skin and mucosae. Not only are they, generally speaking, harmless, but they can also be protective for the host. Most of the bacteria in the porcine alimentary tract, for instance, are commensals. Some bacteria belonging to the microbiota may cause disease under certain circumstances, for instance, when the host's immunity is compromised in some way or when the normal microbiota is disturbed. They are sometimes referred to as facultative pathogens. Actually, most bacteria that cause disease in pigs belong to this group of bacteria. *Trueperella pyogenes*, *Pasteurella multocida*, *Streptococcus suis*, *Escherichia coli*, *Fusobacterium necrophorum*, *Clostridium perfringens* type A (*C. perfringens*), and *Glaesserella parasuis* are examples of such bacteria. Obligate pathogens, on the other hand, do not belong to the normal microbiota. This terminology is sometimes also used to refer to bacteria that may cause disease in healthy, immunocompetent hosts, such as *Bacillus anthracis*. Opportunistic pathogenic bacteria generally exhibit low virulence, but if the host becomes injured or is seriously immunocompromised, they may cause disease.

Pathogenicity refers to the potential ability of an organism to cause disease and harm the host. This ability represents a genetic component of the pathogen, and the overt damage done to the host is a property of the host–pathogen interactions (Thomas and Elkinton, 2004). Mutualistic and most commensal bacteria belonging to the normal microbiota lack this inherent ability to cause disease. However, disease is not an inevitable outcome of the host–pathogen interaction and, furthermore, pathogens can express a wide range of virulence. Host immune factors, the number of organisms present in the initial exposure, and the virulence attributes of a bacterium all play a role in the development of disease. Virulence, a term often used interchangeably with pathogenicity, refers to the degree of disease caused by the organism. The extent of the virulence is usually correlated with the ability of the pathogen to multiply within the host and may be affected by other factors. Pathogenicity is a qualitative term, an 'all-or-none' concept, whereas virulence is a term that quantifies pathogenicity. An organism (species or strain) is defined as being pathogenic (or not) and, depending upon conditions, may exhibit different levels of virulence.

Bacteria may cause disease by different mechanisms; two important ones are tissue invasion and toxin production (Songer and Post, 2005). Tissue invasion can be accomplished by adhering to and/or penetrating cells, producing extracellular substances to facilitate invasion, and overcoming host defenses. Adhesins are surface proteins that facilitate adhesion to host cells. Many strains of pathogenic *E. coli* possess such surface adhesins. *Listeria monocytogenes*, *Mycobacterium* species, *Brucella* species, *Rhodococcus equi*, and *Salmonella enterica* are examples of facultatively intracellular bacteria that are able to invade and multiply in host cells (Niemann et al., 2004). Some bacteria produce extracellular enzymes such as coagulase produced by *Staphylococcus hyicus* and streptokinase produced by beta-hemolytic streptococci, enabling them to spread within the host tissue. Other bacteria mainly cause disease through toxin production. Exotoxins are metabolites that are produced in the bacterial cytoplasm and are mostly released into the extracellular environment by active secretion. Some exotoxins are only released in the environment after the lysis of the bacterial cell. Most exotoxins are proteins that are highly antigenic, and it may be useful to include inactivated exotoxin (toxoid) or a nontoxic but antigenic recombinant protein derived from the exotoxin in vaccines. The thermostable enterotoxins produced by some enterotoxigenic *E. coli* (ETEC) strains are polypeptides that are poorly immunogenic (Haesebrouck et al.,

2004). Exotoxins vary greatly in their potency, ranging from the highly toxic botulinum toxin, the cytolysins (Apx toxins) of *Actinobacillus pleuropneumoniae* to the weakly toxic product released by *T. pyogenes*. Other swine pathogens that may produce exotoxins are *C. perfringens*, enterotoxigenic strains of *E. coli*, *Pasteurella multocida*, and *Staphylococcus hyicus*. Endotoxins are the lipopolysaccharides found in the cell wall of gram-negative bacteria. These may be released from both actively growing bacterial cells and dead bacteria. The release of endotoxins is very important in the pathogenesis of gram-negative bacteria and is partially responsible for the clinical signs produced by these pathogens. Some cell wall components of Gram-positive bacteria, such as lipoteichoic acids and peptidoglycan, can cause similar host reactions as lipopolysaccharides from gram-negative bacteria (Middelveld and Alving, 2000).

Under field conditions, mixed infections with different pathogens (e.g. viruses, bacteria, and parasites) are common. A typical example is the PRDC. An overview of the interactions based on experimental infections with two respiratory pathogens (i.e. either two viruses, one virus and one bacterium, or two bacteria) has been published by Opriessnig et al. (2011). Although the different pathogens are capable of inducing disease independently, simultaneous infection with two or more pathogens often results in more severe disease. Some known mechanisms by which pig pathogens interact include damage to the muco-ciliated epithelium of the respiratory tract, inducing immune suppression, altering cytokine responses, or affecting macrophage function (Opriessnig et al., 2011). Also the sequence of infection with different pathogens likely plays a role in disease outcome.

5 Transmission of swine bacterial pathogens

The different routes of transmission of a selected number of important swine bacterial pathogens are shown in Table 2. The table is based on Amass (2005) and information from the literature published since then on the transmission of these pathogens has been added.

Transmission by direct contact occurs for all bacterial pathogens, whereas indirect transmission may vary depending on the pathogen. The indirect routes have been classified arbitrarily in the following categories: people, semen, manure, domestic/feral animals, rodents, insects (vectors), aerosol, animal feed, water, and fomites. Indirect transmission by close contact with infected droplets or feces and mechanical transfer by fomites or vectors are the ways in which bacteria are commonly spread.

Most bacterial pathogens can be transmitted from the sow to the offspring. For some pathogens like *Streptococcus suis*, piglets may be colonized already

Table 2 Direct and indirect transmission routes of swine pathogens and diseases

Disease/pathogen	References	Direct contact — Purchasing policy, Transport/removal of animals/manure/cadavers, all subcategories of internal biosecurity[a]	People — Access check indirectly: transport/removal of animals/manure/cadavers; supply of fodder, water, and equipment; cleaning and disinfection	Semen — Purchasing policy
Actinobacillus pleuropneumoniae	Assavacheep and Rycroft (2013); Loera-Muro et al. (2013); Tobias et al. (2014); Baroch et al. (2015); Marinou et al. (2015); Gottschalk and Broes (2019)	O		
Bordetella bronchiseptica	Nicholson et al. (2012); Brockmeier et al. (2019)	O		
Brachyspira hyodysenteriae	Desrosiers (2011); Hampson and Burrough (2019); Zeeh et al. (2020)	O	O	
Brucella suis	Wu et al. (2012); Lama and Bachoon (2018); Olsen et al. (2019)	O	X	X
Clostridium perfringens	Allaart et al. (2013); Uzal and Songer (2019)	O		
Erysipelothrix rhusiopathiae[c]	Bender et al. (2010); Boadella et al. (2012); Malmsten et al. (2018); Marinou et al. (2015); Opriessnig and Coutinho (2019)	O		
Escherichia coli	Cornick and VuKhac (2008); Callens et al. (2015); Pearson et al. (2016); Lama and Bachoon (2018); Fairbrother and Nadeau (2019)	O	X	
Glaesserella parasuis[1]	Brockmeier et al. (2013); Aragon et al. (2019)	O		
Lawsonia intracellularis[1]	Friedman et al. (2008); Baroch et al. (2015); Pearson et al. (2016); Vannucci et al. (2019)	O		
Leptospires	Maes et al. (2008); Althouse and Rossow (2011); Arent and Ellis (2019); Cilia et al. (2020)	O	X	O
Mycoplasma hyopneumoniae	Nathues et al. (2012); Baroch et al. (2015); Marinou et al. (2015); Pearson et al. (2016); Malmsten et al. (2018); Pieters and Maes (2019)	O	O	
Pasteurella multocida	Fablet et al. (2011); Desrosiers (2011); Register and Brockmeier (2019)	O	O	
Salmonella enterica.	Andres and Davies (2015); Martelli et al. (2018); Griffith et al. (2019)	O	X	
Streptococcus suis	Fablet et al. (2011); Gottschalk and Segura (2019); Giang et al. (2020).	O	X	

[a] As described in Biocheck.UgentTM.
[b] Transmitted to distances >2 km.
[c] Added to the original list of Amass (2005). 'X' represents possible transmission routes described by Amass (2005). 'O' describes new possible transmission routes based on studies published between 2005 and January 2021.
Source: Adapted from Amass (2005) and Filippitzi et al. (2018).

	Indirect contact						
Manure	Domestic/ feral animals	Rodents	Insects (vectors)	Aerosol	Animal feed	Water	Fomites
Transport/ removal of animals/ manure/ cadavers	Vermin and bird control	Vermin and bird control	Location, environment	Purchasing policy, transport/ removal of animals/ manure/cad avers, loca tion, envir onment, All	Supply of fodder, water and equipment	Supply of fodder, water and equipment	Cleaning and disinfection, indirectly: transport/removal of animals/manure/ cadavers; supply of fodder, water, and equipment; Access check; Vermin and bird control
	O			X		O	O
	X	X	O	X		X	O
X	X	X	O		O	X	O
O	X		O	O	O		
O			O	X		X	O
O	O	O			O	O	O
X	O	X	X	X	O	X	X
	O						
O	O	O	O				O
	X	X				X	
	O			X^b		X	X
X	O			X		X	O
X	X	X	X	X	X	X	X
X	X		X	X		X	X

during or shortly after birth from vaginal fluids or while nursing. For other pathogens such as *Mycoplasma hyopneumoniae* and *A. pleuropneumoniae*, transmission from sow to offspring takes place later during lactation. For the latter pathogens, strategies such as (medicated) early weaning can be used to obtain piglets that are free from pathogens that are endemic in the sow population (Alexander et al., 1980).

6 Clinical signs and lesions induced by swine bacterial pathogens

An overview of the clinical signs and gross postmortem lesions of gram-positive and gram-negative swine bacterial pathogens are shown in Tables 3 and 4 (Muirhead and Alexander, 1997; Post, 2019).

A particular situation is the presence of bacteria in the ejaculates of breeding boars. Apart from the risk for the transmission of bacterial pathogens like *Brucella suis* to recipient sows inseminated with contaminated semen, bacterial contamination is also known to be detrimental to semen quality as it will cause sperm agglutination and reduce motility (Althouse and Lu, 2005). It may also decrease the longevity of the sperm during storage and the fertility potential (Sepúlveda et al., 2013, 2014).

Commonly isolated bacteria from extended semen and their effect on sperm quality are summarized in Table 5. They mostly belong to the Enterobacteriaceae family (Althouse and Lu, 2005; Úbeda et al., 2013). Within this family, *Serratia marcescens*, *Klebsiella* species, *Morganella* species, or *Proteus* species have been demonstrated to be present in a high percentage of samples and their presence is associated with reduced motility (Úbeda et al., 2013). *Enterobacter cloacae* at sperm:bacteria ratio of 1:5 and 1:10 reduced sperm motility and membrane integrity and resulted in sperm agglutination in semen doses stored at 15-17°C (Prieto-Martinez et al., 2014). *Clostridium perfringens* reduced sperm motility and viability after the inoculation of 10^8 cfu ml^{-1} in semen doses and 24 h incubation at 37°C or storage at 15°C (Sepúlveda et al., 2013). Similarly, experimental contamination of stored boar semen with *Pseudomonas aeruginosa* resulted in significant decreases in the percentages of sperm motility, viability, and acrosome integrity but did not affect pH (Sepúlveda et al., 2014). In a study by Althouse et al. (2000), extended semen samples were inoculated with colonies of the six most frequently isolated bacteria (*Enterobacter cloacae*, *E. coli*, *Serratia marcescens*, *Alcaligenes xylosoxidans*, *Burkholderia cepacia*, and *Stenotrophomonas maltophilia*). For all isolates, visual clumping, microscopic sperm-to-sperm agglutination, poor motility, and damaged acrosomes were observed after inoculation in a time-dependent manner. The exact interactions

Table 3 Gram-positive bacteria in swine and associated disease(s) and/or clinical signs

Bacterium	Disease(s)/clinical signs/lesions
Actinobaculum suis	Cystitis, pyelonephritis
Actinomyces hyovaginalis	Sporadic abortion, embolic pneumonia
Bacillus anthracis	Anthrax (pharyngeal, intestinal, septicemic form)
Clostridium botulinum	Botulism (paralysis, death) (pigs are not very sensitive)
Clostridium chauvoei	Blackleg (emphysematous necrotizing myositis) (pigs are not very sensitive)
Clostridioides (Clostridium) difficile	Neonatal colitis
Clostridium perfringens	Neonatal enteritis, pseudomembranous colitis
Clostridium novyi	Sudden death, hepatitis
Clostridium septicum	Malignant edema
Clostridium tetani	Tetanus (generalized skeletal muscle spasms)
Enterococcus sp.	Enteritis
Erysipelothrix rhusiopathiae	Erysipelas (acute, subacute, chronic form)
Listeria monocytogenes	Listeriosis, abortion, encephalitis, septicemia
Mycobacterium sp.	Tuberculosis (local, generalized form)
Mycoplasma hyopneumoniae	Enzootic pneumonia (broncho-interstitial pneumonia)
Mycoplasma hyorhinis	Arthritis, otitis, polyserositis
Mycoplasma hyosynoviae	Arthritis
Mycoplasma suis	Anemia, infertility, poor growth, pericarditis, unthriftiness
Rhodococcus equi	Granulomatous lymphadenitis
Staphylococcus aureus	Abscesses, arthritis, enteritis, mastitis, metritis, neonatal septicemia, vaginitis
Staphylococcus hyicus	Exudative epidermitis (greasy pig disease)
Streptococcus dysgalactiae spp. *equisimilis*	Arthritis, endocarditis, meningitis, septicemia
Streptococcus porcinus	Cervical lymphadenitis
Streptococcus suis	Septicemia, arthritis, endocarditis, meningitis, polyserositis
Trueperella abortisuis	Sporadic abortion
Trueperella pyogenes	Abscesses, arthritis, endocarditis, mastitis, osteomyelitis, pneumonia

Source: Adapted from Muirhead and Alexander (1997) and Post (2019).

of the different bacteria with sperm are not yet fully investigated. It has been suggested that *P. aeruginosa* decreases the ability of sperm to accomplish capacitation (Sepúlveda et al., 2016).

Table 4 Gram-negative bacteria in swine and associated disease(s) and/or clinical signs

Bacterium	Disease(s)/clinical signs/lesions
Actinobacillus pleuropneumoniae	Pleuropneumonia (peracute, acute, chronic form)
Actinobacillus suis	Pneumonia, septicemia
Bordetella bronchiseptica	Nonprogressive atrophic rhinitis, pneumonia
Burkholderia pseudomallei	Melioidosis, internal abscesses, lymph node abscesses
Brachyspira hyodysenteriae, Brachyspira hampsonii, or Brachyspira suanatina	Swine dysentery, mucohemorrhagic enteritis of the large intestine
Brachyspira pilosicoli	Porcine intestinal/colonic spirochetosis
Brachyspira murdochii or Brachyspira intermedia	Mild colitis, loose stools
Brucella suis	Brucellosis, abortion, orchitis, arthritis, infertility
Campylobacter coli/jejuni	Subclinical enterocolitis, mild diarrhea
Chlamydia suis	Conjunctivitis, neonatal diarrhea, pneumonia, possibly abortion
Chlamydia abortus	Early embryonic death, abortion
Escherichia coli	Colibacillosis, edema disease, cystitis, enteritis, mastitis, neonatal septicemia
Glaesserella parasuis	Glässer's disease, arthritis, polyserositis
Helicobacter suis	Gastritis, gastric ulcers
Klebsiella pneumoniae	Piglet septicemia
Lawsonia intracellularis	Porcine proliferative enteropathy
Leptospira sp.	Leptospirosis, infertility, stillbirths, weak piglets
Pasteurella multocida	Progressive atrophic rhinitis, pneumonia
Salmonella enterica	Enteric and septicemic salmonellosis
Treponema pedis	Cutaneous spirochetosis (granulomatous lymphadenitis)
Yersinia enterocolitica	Subclinical enterocolitis
Yersinia pseudotuberculosis	Diarrhea, enterocolitis

Source: Adapted from Muirhead and Alexander (1997) and Post (2019).

7 Diagnosis of swine bacterial pathogens

Developing an efficient diagnostic approach should start with gathering all relevant farm information and a detailed description of the problem. Further diagnostic workup is mostly necessary to establish a conclusive diagnosis, to elucidate the pathogens involved, and to assess their relative impact on the problem.

Table 5 Percentage of contaminated extended semen samples[a] in which different bacteria were isolated and their effect on sperm quality

Bacteria	Althouse and Lu (2005)	Schulze et al. (2015)	Úbeda et al. (2013)	Effect on sperm
Total contaminated samples	31.2% (78/250)	25.6% (88/344)	14.7% (263/1785)	Reduced sperm quality
Achromobacter xylosoxidans	10.3%	3.4%	ND	Agglutination, poor motility, damaged acrosomes, acidic pH (Althouse et al., 2000)
Burkholderia cepacia	2.6%	ND	ND	Agglutination, poor motility, damaged acrosomes, acidic pH (Althouse et al., 2000)
Clostridium perfringens	ND	ND	ND	Poor sperm viability and motility (Sepúlveda et al., 2013)
Enterobacter cloacae	2.6%	13.6%	ND	Agglutination, poor motility, damaged acrosomes, acidic pH, decreased the osmotic resistance (Althouse et al., 2000; Prieto-Martinez et al., 2014)
Enterococcus spp.	20.5%	8%	ND	ND
Escherichia coli	6.4%	ND	1.5%	Agglutination, poor motility, damaged acrosomes (Althouse et al., 2000)
Klebsiella spp.	3.8%	8%	11.8%	Poor motility (Úbeda et al., 2013)
Leifsonia aquatica	ND	20.5%	ND	ND
Morganella morganii	ND	ND	3.8%	Poor motility, damage acrosome, poorer response to the hypoosmoic swelling test (Úbeda et al., 2013)
Proteus mirabilis	1.3%	5.7%	1.9%	Poor motility, abnormal forms (Úbeda et al., 2013)
Providencia spp.	3.8%	ND	9.1%	ND
Pseudomonas spp.	6.4%	5.7%	ND	*P. aeruginosa* reduced total and progressive sperm motility, sperm viability, and acrosome integrity
Ralstonia pickettii	ND	11.4%	ND	ND
Stenotrophomonas maltophilia	15.4%	ND	ND	Agglutination, poor motility, damaged acrosomes
Serratia marcescens	10.3%	2.3%	12.5%	Agglutination, poor motility, damaged acrosomes, acidic pH (Althouse et al., 2000)

[a]The percentage refers to the total contaminated samples. Bacteria present in a percentage of samples lower than 5% and identified only in one of the studies and for which no effect on sperm quality has been described are not included in the table. ND = not described. Source: Adapted from Lopéz et al. (2017).

Necropsies of a representative number of typically affected animals can be very helpful. The basic lesion caused by *Mycoplasma hyopneumoniae* and swine influenza viruses is a broncho-interstitial pneumonia; subsequent infection by other respiratory bacteria ends up with suppurative bronchopneumonia if the damage does not affect pleura and pleuropneumonia if it affects the pleura. It must be reminded that infections by *Glaesserella parasuis* and *Streptococcus suis* are systemic, and pleuritis usually occurs in the absence of pneumonia (Maes et al., 2001). However, mixed infections are generally the rule, and lesions may not appear very typical for one specific pathogen. Performing necropsies is mostly followed by taking samples for further diagnostic examinations such as histopathology, immunohistochemistry, pathogen isolation, or PCR testing.

In addition to investigating dead or euthanized animals, living animals from the affected age group can be sampled to detect antibodies and/or possible pathogens involved. Different sampling strategies can be used, depending on the age group, the type of sample, the disease, and the diagnostic test used. In the case of serology, paired serum samples will often be needed to establish the diagnosis. To save analysis costs, individual samples like fecal samples for *Brachyspira hyodysenteriae* may be pooled (Neirynck et al., 2020). For detection of *Salmonella* in subclinically infected pigs, overshoe samples may be used. They are collected at pen or compartment level and are more likely to be *Salmonella* positive than when samples from a limited number of individual animals are collected as the shedding of the bacteria in the feces by subclinically infected pigs is intermittent (Peeters et al., 2019). A detailed overview of sample selection, collection, and submission for different swine pathogens has been published by Arruda and Gauger (2019).

Examination of pigs in the slaughterhouse is an additional and valuable tool to obtain information on the health of the animals in an easy and cheap way (Alban et al., 2015). Different lesions can be monitored like lesions in the respiratory, gastrointestinal, or urogenital tract; tail biting lesions; joint lesions; and abscesses (Maes et al., 2001; de Jong et al., 2014). If the lesions are monitored from different successive batches of a herd, time evolution can be assessed. Slaughter checks are also useful to detect subclinical infections that produce gross lesions. Most scoring methods are based on a visual subjective estimation of the proportion of affected tissue. Therefore, it can involve possible errors. In addition, most lesions are not pathognomonic for a specific pathogen.

8 Control of swine bacterial pathogens

8.1 General control measures

Comprehensive reviews of the different control measures for particular swine bacterial pathogens have been published elsewhere (Luppi, 2017; Marco et al., 2020). They will not be repeated in detail here, even though they are of critical importance. Factors that should be considered to improve the overall biosecurity in pig farms are presented in a separate chapter in this book. A short overview of the most important control measures for swine bacterial disease will be discussed here. The measures are classified into the following categories: (1) type of production system, (2) replacement rates and gilt acclimation, (3) management, (4) hygiene, (5) nutrition, and (6) climate and housing conditions.

8.1.1 Type of production system

Control factors related to the type of production system include piglet source (one source lower risk than multiple sources), pig flow (not mixing of age groups) and batch system (more than 1-week systems lower risk than 1-week system), and parity segregation (separation of primiparous and multiparous sows). Closed herds or closed production systems that are geographically isolated and do not purchase pigs have the advantage that there is no risk for the introduction of bacterial pathogens via the introduction of new animals. However, in case closed farms are endemically infected, the infection is maintained in the farm, and breeding animals will continue to transmit the pathogen to the offspring. In case of multisite production systems, there is a risk to contract the disease by moving (subclinically) infected animals between the sites.

8.1.2 Replacement rates and gilt acclimation

Low replacement rates, purchasing breeding gilts from herds with similar or better health status, quarantine and implementing proper gilt acclimation practices can lower the risk of introducing a pathogenic bacterial species into the farm. As animals are often subclinically infected with bacterial pathogens (such as *Mycoplasma hyopneumoniae*, *A. pleuropneumoniae*, *Salmonella* Typhimurium, *Brachyspira hyodysenteriae*, and *Streptococcus suis*), the introduction of clinically healthy but infected pigs represents a great risk. Therefore, information from the origin farm, quarantine, and proper testing are needed.

Replacement gilts negative for specific bacterial pathogens have become available from breeding stock companies. However, also the introduction of naïve gilts into endemically infected farms may be a challenge. Naïve gilts will be exposed to positive sows and may eventually get infected and shed large numbers of bacteria. These gilts also become an infectious reservoir for newborn piglets. In addition, females of the recipient herd may be exposed to recirculation of the bacterium and become reinfected, generating an unstable disease status in the herd.

Proper gilt acclimation procedures, such as vaccination during the quarantine period, are therefore needed. Gilt acclimation will enhance immunity and alleviate or prevent introduction problems when gilts join an endemically infected breeding population. In the case of *Mycoplasma hyopneumoniae*, strategies have been used to purposefully expose incoming gilts to *Mycoplasma hyopneumoniae* at an early age, with the aim to immunize gilts and prevent shedding prior to entering the sow farm (Pieters and Fano, 2016).

8.1.3 Management

Management practices that are beneficial against bacterial disease include measures prior to weaning (sufficient colostrum intake, limiting cross-fostering, increased weaning age), strict all-in/all-out production, reducing stress rates (by maintaining optimal stocking densities and group size), prevention of other pathogens including viruses and parasites, and control of mycotoxins (Rhouma et al., 2017; Marco et al., 2020). Management strategies like weaning at a later age (28–33 instead of 21 days) ameliorate the stress of the piglets and reduce the risk for disease problems post-weaning (Faccin et al., 2020). However, more research is needed in this area including effects of changes in facilities like comingling systems, group lactations, or supplementary milk systems. All-in/all-out management with cleaning and disinfection between batches reduces the transmission of infection. It is of special importance to avoid any pig being delayed in the flow. All those pigs suffering the disease should be euthanized or moved forward with their group to avoid affecting the next batch, especially during an active outbreak. Stressful conditions such as pig handling, crowding, transportation, or severe weather conditions should be minimized to prevent clinical outbreaks in herds at risk.

8.1.4 Hygiene

A thorough cleaning to remove organic matter and disinfection of the pens should be practiced to limit or prevent carryover of pathogens from a previous batch of pigs to the next batch in the same barn (Dutkiewicz et al., 2017). In case of problems with a particular disease, a disinfectant that is active against the pathogen should be selected. Keeping the barns as dry as possible after disinfection may further lower the infection levels and minimize the effects of the disease. Some bacterial pathogens like *Brachyspira hyodysenteriae* and *Salmonella* Typhimurium can be transmitted via rodents, domestic animals (e.g. cats and dogs), birds, and insects. Mice and rats can serve as potential reservoirs of these pathogens in pig herds, so implementation of effective rodent control is essential. Dogs and cats should not be allowed to have access to the pig stables. Also, stables should be preferably bird proof, so that birds do not have access to the area where the pigs are housed. Contact with wild boar or foxes should be avoided (Zeeh et al., 2020). Proper insect control is also needed to control the transmission of many bacterial pathogens such as *Brachyspira hyodysenteriae*, *Mycoplasma suis*, and *Streptococcus suis*. Unfortunately, in the case of outdoor pig production, it is almost impossible to prevent mechanical transmission of infectious material via birds, waterfowl, and other potential wildlife vectors. Peat used as bedding material for piglets might be contaminated by *Mycobacterium avium* and lead to infections in pigs (Johansen et al., 2014). Therefore, potentially contaminated bedding material should not be used or should be properly decontaminated.

8.1.5 Nutrition

Attention should also be paid to the composition and form of the diet as a clinical expression of enteric infections like *E. coli* infections post-weaning, *Brachyspira hyodysenteriae* and *Salmonella* infections is often influenced by dietary factors. Diet can influence the pH and the composition of the microbiota in the intestinal tract. Some diet components such as antimicrobial peptides, chitosan, lysozyme, short- and medium-chain fatty acids/triglycerides and plant extracts can have direct antibacterial effects as well (Vanrolleghem et al., 2019). Good quality ingredients, milk products, and digestible diets with a low protein content reduce the risk for post-weaning diarrhea (Luppi, 2017). Zinc oxide at high concentrations has been shown to be effective in controlling *E. coli* infections post-weaning. Its use will be banned in the EU from 2022 onward (EMA, 2017) because of the environmental concerns and because it has been associated with selection for zinc resistance, and other heavy metals like copper, and methicillin resistance in *Staphylococcus aureus* (Luppi, 2017). Zinc oxide is still commonly used in other parts of the world. Diets that reduce the amount of fermentable substrate entering the large intestine are beneficial against swine dysentery (Leser et al., 2000). The use of a coarse non-pelleted feed is associated with reduced *Salmonella* prevalence (O'Connor et al., 2008). Further research is warranted to assess the efficacy of nutritional supplements such as phytogenic ingredients, essential oils, prebiotics, probiotics, gut acidifiers, and mannan-oligosaccharides. Finally, also good water quality is important for pig health.

8.1.6 Climate and housing conditions

Optimal housing and climatic conditions are important for all bacterial infections but, in particular, for respiratory infections (Marco et al., 2020). Ventilation and hygiene are needed to maintain appropriate environmental thermal conditions in the stables (e.g. temperature, humidity, and airspeed) with minimal fluctuations and minimize air contaminant levels (e.g. dust, ammonia, and endotoxins). Season has an effect on respiratory disease, and generally there are less problems in summer when temperatures are higher and relative humidity is lower.

8.2 Antimicrobial medication

Antimicrobials are necessary for the treatment of bacterial infections and by doing so, we can safeguard animals. They should, however, be used judiciously and restrictively to avoid or limit the risk for antimicrobial resistance in both pathogens and the normal microbiota. High antimicrobial use and the resulting risk for antimicrobial resistance is an important concern worldwide. Prophylactic use or using antimicrobials as a substitute for improper management should be avoided. In some parts of the world, such as the EU, regulations to control antimicrobial resistance do not allow prophylactic use of antimicrobials (EMA, 2020). Animals that are clinically diseased due to a bacterial infection requiring antimicrobial treatment should preferably be treated by injection since they usually do not eat and might drink less. The treatment should be done following the principles stated in the Guidelines for the Prudent Use of Antimicrobials in Veterinary Medicine (2015/C 299/04). In addition, the use of critically important antimicrobials for human medicine such as fluoroquinolones, third- and fourth-generation cephalosporins, and recently also colistin should be restricted or prohibited. This is particularly important not only for infections with zoonotic bacterial pathogens but also for the treatment of other bacterial infections as there is an increasing resistance of E. coli strains isolated from pigs with diarrhea to colistin (Burow et al., 2019).

Judicious use of antimicrobials in animals in order to limit antimicrobial resistance is also receiving increased attention outside Europe, although there is substantial variation in legislation when viewed at an international level (Bright-Ponte, 2020; Hoque et al., 2020; Léger et al., 2019; Ogwuche et al., 2021). The US Food and Drug Administration (FDA) Center for Veterinary Medicine (CVM) has taken important steps to address potential antimicrobial resistance concerns surrounding the use of antimicrobials in food-producing animals, including changes to the approved use conditions of medically important antimicrobials to support their judicious use in food-producing animals (Bright-Ponte, 2020).

8.3 Vaccination

Improving the immunity status of susceptible animals by vaccination is a helpful tool to control infectious diseases, including swine bacterial disease. Vaccination is generally able to reduce the (risk for) clinical signs, lesions and performance losses due to disease. However, most vaccines only provide partial protection, do not prevent infection, and are not able to eliminate the pathogens from the herd. The decision of whether to implement vaccination is not straightforward. It depends on the pig producer (degree of being 'risk averse'), the type and severity of disease, the available vaccines and expected effects, and other factors such as the type of herd, flow of the pigs, infection level, and management and housing conditions. Bacterial diseases for which commercial vaccines are available are shown in Table 6. Some vaccines contain the inactivated toxin or a nontoxic but antigenic recombinant protein derived from the exotoxin. Examples of this group include vaccines against *A. pleuropneumoniae*, clostridial infections, enterotoxigenic *E. coli*, shiga-toxin-producing *E. coli* and dermonecrotic toxin *Pasteurella multocida*, and *Bordetella bronchiseptica* (progressive atrophic rhinitis). Other vaccines target extracellular bacteria, and protection is generally mediated by antibodies against their surface antigens and certain secreted antigens, but cell-mediated immunity may also be important. Examples of this group are vaccines against *Erysipelothrix rhusiopathiae* (*E. rhusiopathiae*) and *Mycoplasma hyopneumoniae*. Finally, there are also vaccines against (facultative) intracellular bacteria such as *L. intracellularis* and *Salmonella* Typhimurium. For protection against these bacteria, cell-mediated immune responses are important (Haesebrouck et al., 2004).

Vaccination schemes are highly variable between herds and also between countries. Gilts in the quarantine are often vaccinated against multiple bacterial pathogens (*A. pleuropneumoniae*, dermonecrotic toxin *Pasteurella multocida* and *Bordetella bronchiseptica*, *Clostridium* spp., *E. coli*, *E. rhusiopathiae*, *G. parasuis*, *Leptospira* spp., *L. intracellularis*, and *Mycoplasma hyopneumoniae*) as part of the acclimation practices. Breeding sows are often vaccinated against neonatal *E. coli* and clostridium infections and *E. rhusiopathiae*, and in some herds also against other bacterial pathogens such as dermonecrotic toxin *Pasteurella multocida* and *Bordetella bronchiseptica*, *G. parasuis*, *A. pleuropneumoniae*, and *Mycoplasma hyopneumoniae* infections, leptospirosis, salmonellosis, and so on. Some vaccinations in sows aim to provide protection to the piglets via maternally derived antibodies. Typical examples are neonatal *E. coli* and *Clostridium* infections and infections with dermonecrotic toxin *Pasteurella multocida* and *Bordetella bronchiseptica*. For these vaccinations to be successful, piglets should have sufficient colostrum intake (Declerck et al., 2015). Vaccination against *Mycoplasma hyopneumoniae* in piglets is very frequently practiced worldwide. Other bacterial vaccines that are also available for piglets

Table 6 Common bacterial infections and diseases in the pig and availability of commercial vaccines

	Bacterium		Disease/clinical sign/lesion	Age group	Commercial vaccine[a]
Enteric	Escherichia coli (ETEC, EPEC, STEC)[b]		Neonatal diarrhea	1-3 days	Yes
			Pre-weaning diarrhea	7-21 days	No
			Post-weaning diarrhea	5-14 days post weaning	Yes
			Edema disease	5-14 days post weaning	Yes
	Clostridium perfringens	Type C	Necrotic enteritis	1-14 days	Yes
		Type A	Diarrhea	3-28 days, weaned pigs	Yes
	Clostridioides (Clostridium) difficile		Diarrhea, ill thrift	5-21 days, rarely older	No
	Helicobacter suis		Gastritis, gastric ulcers	Fattening pigs, adults	No
	Salmonella enterica	Typhimurium	Occasional diarrhea, septicemia, death	All ages, usually after weaning	Yes
		Derby			No
		Choleraesuis	Occasional diarrhea Septicemia, diarrhea, death		Yes
	Lawsonia intracellularis[c]		Porcine proliferative enteropathy (ileitis)	Nursery-fattening pigs	Yes[c]
			Regional/necrotic ileitis	Nursery-fattening pigs	
			Porcine hemorrhagic enteropathy	Fattening pigs, young adults	
	Strongly hemolytic Brachyspira spp.		Swine dysentery	All ages, mainly grower-fattening pigs	No
	Brachyspira pilosicoli		Porcine intestinal spirochaetosis (PIS)	5 weeks to 4 months	No
Respiratory	Dermonecrotic toxin Pasteurella multocida (D) Bordetella bronchiseptica		Progressive atrophic rhinitis (PAR)	1-8 weeks Nasal distortion lasts for life	Yes
	Mycoplasma hyopneumoniae		Enzootic pneumonia	Grower-fattening pigs	Yes
	Pasteurella multocida		Mycoplasma-induced respiratory disease	Grower-fattening pigs - secondary invader	No

	Actinobacillus pleuropneumoniae	Pleuropneumonia	Grower-fattening pigs	Yes
Septicemic/ bacteremic/ other infections	E. coli	Bacteremia, arthritis, navel infections Cystitis, nephritis	Pigs post-weaning Sows	No
	Streptococcus suis	Meningitis, endocarditis, arthritis, and peritonitis	2–10 weeks	No
	Glaesserella parasuis	Glässer's disease	2–10 weeks	Yes
	Mycoplasma hyosynoviae	Mycoplasmal arthritis	>16 weeks	No
	Staphylococcus aureus	Bacteremia, arthritis, osteomyelitis, mastitis and metritis	All age groups	No
	Staphylococcus hyicus	Exudative epidermitis	Pre- and post-weaning piglets	No
	Erysipelothrix rhusiopathiae	Erysipelas (dermatitis, arthritis, endocarditis)	Grower-fattening pigs, breeding pigs	Yes
	Leptospira spp.	Leptospirosis	Breeding animals	Yes

a Commercial vaccines available in only one or a few countries are not included in the table.
b ETEC: enterotoxic; EPEC enteropathogenic; STEC Shiga-toxin-producing E. coli.
c Infections with Lawsonia intracellularis may lead to different pathological/clinical forms depending on the age of the animals and other factors. Commercial vaccines are available against L. intracellularis. Source: Adapted from Aarestrup et al. (2008) and Thomson and Friendship (2019).

include *A. pleuropneumoniae*, *G. parasuis*, post-weaning enterotoxigenic *E. coli* and shiga-toxin-producing *E. coli*, *L. intracellularis*, *Salmonella* spp., and so on.

In general, vaccination is economically justified in clinically affected herds, especially if no major risk factors can be identified or changed in short term. However, vaccination may also be cost-efficient in subclinically infected herds as subclinical infections with pathogens like *Mycoplasma hyopneumoniae* or *A. pleuropneumoniae* also reduce performance (Regula et al., 2000; Maes et al., 2003). For some important pig pathogens (e.g. *Brachyspira hyodysenteriae* and *Streptococcus suis*), however, no or few commercial vaccines are available. Autogenous vaccines may also be used. These are mostly inactivated vaccines, bacterins based on the pathogen strain(s) isolated from animals affected by the disease. Autogenous vaccines against different pathogens such as *Streptococcus suis*, *A. pleuropneumoniae*, and *E. coli* have been described (Baums et al., 2010). One of the disadvantages of autogenous vaccines is that often limited data are available on vaccine efficacy and safety (Haesebrouck et al., 2004).

9 Conclusions

Infections with swine bacterial pathogens significantly affect the health, well-being, and performance of pigs in all types of production systems worldwide, and this will likely remain in the future. The pathogens can be classified according to different characteristics; some of them may determine their interaction with the host and virulence and/or may be important for the epidemiology, diagnostics, treatment, and control measures. Control should focus primarily on the improvement of management, nutrition and housing conditions, and also on vaccination. Most of these measures provide only partial protection against clinical signs, lesions, and production losses. Advances in research on pathogen characteristics, epidemiology, and pathogenesis may improve diagnostics and contribute to the development of new vaccines that provide better protection.

10 Where to look for further information

Diseases of Swine, Eleventh Edition. Editors: Jeffrey J. Zimmerman, Locke A. Karriker, Alejandro Ramirez, Kent J. Schwartz, Gregory W. Stevenson, and Jianqiang Zhang. Published 2019 by John Wiley & Sons, Inc.

Porcine Health Management: The Official Journal of the European College of Porcine Health Management and the European Association of Porcine Health Management: https://porcinehealthmanagement.biomedcentral.com/.

Journal of Swine Health and Production: The Official Journal of the American Association of Swine Veterinarians: https://www.aasv.org/shap.html.

Book Mycoplasmas in Swine. Editors: Dominiek Maes, Marina Sibila, Maria Pieters. Acco, Leuven Belgium, pp. 344.

Advancements and Technologies in Pig and Poultry Bacterial Disease Control. Editors: Neil Foster, Ilias Kyriazakis, Paul Burrow. Elsevier, ISBN 978-0-12-818030-3, pp 278.

Achieving Sustainable Production of Pig Meat. Volume 3: Animal Health and Welfare. Editors: Julian Wiseman. Burleigh Dodds Science Publishing, Boek Wiseman, 328 pp.

Biosecurity in Animal Production and Veterinary Medicine: From Principles to Practice. Editors: Jeroen Dewulf, Filip Van Immerseel. Acco Leuven, 524 pp.

11 References

Aarestrup, F. M., Oliver Duran, C. and Burch, D. G. (2008). Antimicrobial resistance in swine production. *Anim. Health Res. Rev.* 9(2), 135-148.

Alban, L., Petersen, J. V. and Busch, M. E. (2015). A comparison between lesions found during meat inspection of finishing pigs raised under organic/free-range conditions and conventional, indoor conditions. *Porcine Health Manag.* 1, 4.

Alexander, T. J., Thornton, K., Boon, G., Lysons, R. J. and Gush, A. F. (1980). Medicated early weaning to obtain pigs free from pathogens endemic in the herd of origin. *Vet. Rec.* 106(6), 114-119.

Allaart, J. G., van Asten, A. J. A. M. and Grone,A. (2013). Predisposing factors and prevention of *Clostridium perfringens*-associated enteritis. *Comp. Immunol. Microbiol. Infect. Dis.* 36(5), 449-464.

Althouse, G. C., Kuster, C. E., Clark, S. G. and Weisiger, R. M. (2000). Field investigations of bacterial contaminants and their effects on extended porcine semen. *Theriogenology* 53(5), 1167-1176.

Althouse, G. C. and Lu, K. G. (2005). Bacteriospermia in extended porcine semen. *Theriogenology* 63(2), 573-584.

Althouse, G. C. and Rossow, K. (2011). The potential risk of infectious disease dissemination via artificial insemination in swine. *Reprod. Domest. Anim.* 46(Suppl. 2), 64-67.

Amass, S. F. (2005). Biosecurity: Stopping the bugs from getting in. *Pig J.* 55, 104-114.

Andres, V. M. and Davies, R. H. (2015). Biosecurity measures to control salmonella and other infectious agents in pig farms: A review. *Compr. Rev. Food Sci. Food Saf.* 14(4), 317-335.

Aragon, V., Segalés, J. and Tucker, D. (2019). Chapter 54: Glässer's disease. In: Zimmerman, J. J., Karriker, L. A., Ramirez, A., Schwartz, K. J., Stevenson, G. W. and Zhang, J. (Eds) *Diseases of Swine* (11th edn, vol. 2019, pp. 844-853). Hoboken, NJ: John Wiley & Sons, Inc.

Arent, Z. and Ellis, W. (2019). Chapter 55: Leptospirosis. In: Zimmerman, J. J., Karriker, L. A., Ramirez, A., Schwartz, K. J., Stevenson, G. W. and Zhang, J. (Eds) *Diseases of Swine* (11th edn, vol. 2019, pp. 854-862). Hoboken, NJ: John Wiley & Sons, Inc.

Arruda, P. H. E. and Gauger, P. (2019). Chapter 7: Optimizing sample selection, collection and submission to optimize diagnostic value. In: Zimmerman, J. J., Karriker, L. A.,

Ramirez, A., Schwartz, K. J., Stevenson, G. W. and Zhang, J. (Eds) *Diseases of Swine* (11th edn, vol. 2019, pp. 98-111). John Wiley & Sons, Inc.

Assavacheep, P. and Rycroft, A. N. (2013). Survival of *Actinobacilluspleuropneumoniae* outside the pig. *Res. Vet. Sci.* 94(1), 22-26.

Balloux, F., Brønstad Brynildsrud, O., van Dorp, L., Shaw, L. P., Chen, H., Harris, K. A., Wang, H. and Eldholm, V. (2018). From theory to practice: Translating whole-genome sequencing (WGS) into the clinic. *Trends Microbiol.* 26(12), 1035-1048.

Baroch, J. A., Gagnon, C. A., Lacouture, S. and Gottschalk, M. (2015). Exposure of feral swine (*Sus scrofa*) in the United States to selected pathogens. *Can. J. Vet. Res.* 79(1), 74-78.

Baums, C. G., Brüggemann, C., Kock, C., Beineke, A., Waldmann, K. H. and Valentin-Weigand, P. (2010). Immunogenicity of an autogenous *Streptococcussuis* bacterin in preparturient sows and their piglets in relation to protection after weaning. *Clin. Vaccine Immunol.* 17(10), 1589-1597.

Bender, J. S., Shen, H. G., Irwin, C. K., Schwartz, K. J. and Opriessnig, T. (2010). Characterization of *Erysipelothrix* species isolates from clinically affected pigs, environmental samples and vaccine strains from six recent swine erysipelas outbreaks in the United States. *Clin. Vaccine Immunol.* 17(10), 1605-1611.

Boadella, M., Ruiz-Fons, J. F., Vicente, J., Martín, M., Segalés, J. and Gortazar, C. (2012). Seroprevalence evolution of selected pathogens in Iberian wild boar. *Transbound. Emerg. Dis.* 59(5), 395-404.

Bright-Ponte, S. J. (2020). Antimicrobial use data collection in animal agriculture. *Zoonoses Public Health* 67(Suppl. 1), 1-5.

Brockmeier, S., Register, K., Nicholson, T. and Loving, C. (2019). Chapter 49: Bordetellosis. In: Zimmerman, J. J., Karriker, L. A., Ramirez, A., Schwartz, K. J., Stevenson, G. W. and Zhang, J. (Eds) *Diseases of Swine* (11th edn, vol. 2019, pp. 767-777). Hoboken, NJ: John Wiley & Sons, Inc.

Brockmeier, S. L., Loving, C. L., Mullins, M. A., Register, K. B., Nicholson, T. L., Wiseman, B. S., Baker, R. B. and Kehrli, M. E. (2013). Virulence, transmission and heterologous protection of four isolates of *Haemophilus parasuis*. *Clin. Vaccine Immunol.* 20(9), 1466-1472.

Burow, E., Rostalski, A., Harlizius, J., Gangl, A., Simoneit, C., Grobbel, M., Kollas, C., Tenhagen, B. A. and Käsbohrer, A. (2019). Antibiotic resistance in *Escherichia coli* from birth to slaughter and its association with antibiotic treatment. *Prev. Vet. Med.* 165, 52-62.

Callens, B., Faes, C., Maes, D., Catry, B., Boyen, F., Francoys, D., de Jong, E., Haesebrouck, F. and Dewulf, J. (2015). Presence of antimicrobial resistance and antimicrobial use in sows are risk factors for antimicrobial resistance in their offspring. *Microb. Drug Resist.* 21(1), 50-58.

Cilia, G., Bertelloni, F., Piredda, I., Ponti, M. N., Turchi, B., Cantile, C., Parisi, F., Pinzauti, P., Armani, A., Palmas, B., Noworol, M., Cerri, D. and Fratini, F. (2020). Presence of pathogenic Leptospira spp. In the reproductive system and fetuses of wild boars (Sus scrofa) in Italy. *PLoS Negl. Trop. Dis.* 14(12), e0008982.

Cornick, N. A. and VuKhac, H. (2008). Indirect transmission of *Escherichia coli* O157:H7 occurs readily among swine but not among sheep. *Appl. Environ. Microbiol.* 74(8), 2488-2491.

de Jong, E., Appeltant, R., Cools, A., Beek, J., Boyen, F., Chiers, K. and Maes, D. (2014). Slaughterhouse examination of culled sows in commercial pig herds. *Livest. Sci.* 167, 362–369.

Declerck, I., Dewulf, J., Piepers, S., Decaluwé, R. and Maes, D. (2015). Sow and litter factors influencing colostrum yield and nutritional composition. *J. Anim. Sci.* 93(3), 1309–1317.

Desrosiers, R. (2011). Transmission of swine pathogens: Different means, different needs. *Anim. Health Res. Rev.* 12(1), 1–13.

Dutkiewicz, J., Sroka, J., Zając, V., Wasiński, B., Cisak, E., Sawczyn, A., Kloc, A. and Wójcik-Fatla, A. (2017). *Streptococcussuis*: A re-emerging pathogen associated with occupational exposure to pigs or pork products. Part I – Epidemiology. *Ann. Agric. Environ. Med.* 24(4), 683–695.

EMA (European Medicines Agency). (2017). Zinc oxide. Available at: https://www.ema .europa.eu/en/medicines/veterinary/referrals/zinc-oxide. Accessed: 21 March 2022.

EMA (European Medicines Agency). (2020). Categorisation of antibiotics used in animals promotes responsible use to protect public and animal health. Available at: https:// www.ema.europa.eu/en/news/categorisation-antibiotics-used-animals-promotes -responsible-use-protect-public-animal-health. Accessed: 21 March 2022.

European Food Safety Authority and European Centre for Disease Prevention and Control (EFSA and ECDC). (2018). The European Union summary report on trends and sources of zoonoses, zoonotic agents and food-borne outbreaks in 2017. *EFSA J.* 16(12), e05500.

Fablet, C., Marois, C., Kuntz-Simon, G., Rose, N., Dorenlor, V., Eono, F., Eveno, E., Jolly, J. P., Le Devendec, L., Tocqueville, V., Queguiner, S., Gorin, S., Kobisch, M. and Madec, F. (2011). Longitudinal study of respiratory infection patterns of breeding sows in five farrow-to-finish herds. *Vet. Microbiol.* 147(3–4), 329–339.

Faccin, J. E. G., Laskoski, F., Hernig, L. F., Kummer, R., Lima, G. F. R., Orlando, U. A. D., Gonçalves, M. A. D., Mellagi, A. P. G., Ulguim, R. R. and Bortolozzo, F. P. (2020). Impact of increasing weaning age on pig performance and belly nosing prevalence in a commercial multisite production system. *J. Anim. Sci.* 98(4), skaa031.

Fairbrother, J. and Nadeau, E. (2019). Chapter 52: Colibacillosis. In: Zimmerman, J. J., Karriker, L. A., Ramirez, A., Schwartz, K. J., Stevenson, G. W. and Zhang, J. (Eds) *Diseases of Swine* (11th edn, vol. 2019, pp. 807–835). Hoboken, NJ: John Wiley & Sons, Inc.

Filippitzi, M. E., Brinch-Kruse, A., Postma, M., Sarrazin, S., Maes, D., Alban, L., Nielsen, L. R. and Dewulf, J. (2018). Review of transmission routes of 24 infectious diseases preventable by biosecurity measures and comparison of the implementation of these measures in pig herds in six European countries. *Transbound. Emerg. Dis.* 65(2), 381–398.

Friedman, M., Bednár,V., Klimeš,J., Smola, J., Mrlík, V. and Literák, I. (2008). *Lawsoniaintracellularis* in rodents from pig farms with the occurrence of porcine proliferative enteropathy. *Lett. Appl. Microbiol.* 47(2), 117–121.

Giang, E., Hetman, B. M., Sargeant, J. M., Poljak, Z. and Greer, A. L. (2020). Examining the effect of host recruitment rates on the transmission of *Streptococcussuis* in nursery swine populations. *Pathogens* 9(3), 174.

Gottschalk, M. and Broes, A. (2019). Chapter 48: Actinobacillosis. In: Zimmerman, J. J., Karriker, L. A., Ramirez, A., Schwartz, K. J., Stevenson, G. W. and Zhang, J. (Eds) *Diseases of Swine* (11th edn, vol. 2019, pp. 749-766). Hoboken, NJ: John Wiley & Sons, Inc.

Gottschalk, M. and Segura, M. (2019). Chapter 61: Streptococcosis. In: Zimmerman, J. J., Karriker, L. A., Ramirez, A., Schwartz, K. J., Stevenson, G. W. and Zhang, J. (Eds) *Diseases of Swine* (11th edn, vol. 2019, pp. 934-950). Hoboken, NJ: John Wiley & Sons, Inc.

Griffith, R., Carlson, S. and Krull, A. (2019). Chapter 59: Salmonellosis. In: Zimmerman, J. J., Karriker, L. A., Ramirez, A., Schwartz, K. J., Stevenson, G. W. and Zhang, J. (Eds) *Diseases of Swine* (11th edn, vol. 2019, pp. 912-925). Hoboken, NJ: John Wiley & Sons, Inc.

Haesebrouck, F., Pasmans, F., Chiers, K., Maes, D., Ducatelle, R. and Decostere, A. (2004). Efficacy of vaccines against bacterial diseases in swine: What can we expect? *Vet. Microbiol.* 100(3-4), 255-268.

Hampson, D. and Burrough, E. (2019). Chapter 62: Swine dysentery and Brachyspiral colitis. In: Zimmerman, J. J., Karriker, L. A., Ramirez, A., Schwartz, K. J., Stevenson, G. W. and Zhang, J. (Eds) *Diseases of Swine* (11th edn, vol. 2019, p. 951). Hoboken, NJ: John Wiley & Sons, Inc.

Hoque, R., Ahmed, S. M., Naher, N., Islam, M. A., Rousham, E. K., Islam, B. Z. and Hassan, S. (2020). Tackling antimicrobial resistance in Bangladesh: A scoping review of policy and practice in human, animal and environment sectors. *PLoS ONE* 15(1), e0227947.

Johansen, T., Agdestein, A., Lium, B., Jørgensen, J. A. and Djønne, B. (2014). *Mycobacterium aviumsubsp. hominissuis* infection in swine associated with peat used for bedding. *J. Biomed. Biotechnol.* 1, 189649.

Klindworth, A., Pruesse, E., Schweer, T., Peplies, J., Quast, C., Horn, M. and Glöckner, F. O. (2013). Evaluation of general 16S ribosomal RNA gene PCR primers for classical and next-generation sequencing-based diversity studies. *Nucleic Acids Res.* 41(1), e1.

Lama, J. K. and Bachoon, D. S. (2018). Detection of *Brucella suis*, *Campylobacter jejuni*, and *Escherichia coli* strains in feral pig (Sus scrofa) communities of Georgia. *Vector Borne Zoonotic Dis.* 18(7), 350-355.

Lamendella, R., Domingo, J. W., Ghosh, S., Martinson, J. and Oerther, D. B. (2011). Comparative fecal metagenomics unveils unique functional capacity of the swine gut. *BMC Microbiol.* 11, 103.

Léger, A., Alban, L., Veldhuis, A., van Schaik, G. and Stärk, K. D. C. (2019). Comparison of international legislation and standards on veterinary drug residues in food of animal origin. *J. Public Health Policy.* 40(3), 308-341.

Leser, T. D., Lindecrona, R. H., Jensen, T. K., Jensen, B. B. and Moller, K. (2000). Changes in bacterial community structure in the colon of pigs fed different experimental diets and after infection with *Brachyspira hyodysenteriae*. *Appl. Environ. Microbiol.* 66(8), 3290-3296.

Loera-Muro, V. M., Jacques, M., Tremblay, Y. D. N., Avelar-Gonzalez, F. J., Loera-Muro, A., Ramirez-Lopez, E. M., Medina-Figueroa, A., Gonzalez-Reynaga, H. M. and GuerreroBarrera, A. L. (2013). Detection of *Actinobacilluspleuropneumoniae* in drinking water from pig farms. *Microbiology (Reading)* 159(3), 536-544.

López, A., Van Soom, A., Arsenakis, I. and Maes, D. (2017). Boar management and semen handling factors affect the quality of boar extended semen. *Porcine Health Manag.* 3, 15.

Luppi, A. (2017). Swine enteric colibacillosis: Diagnosis, therapy and antimicrobial resistance. *Porcine Health Manag.* 3, 16.

Maes, D., Nauwynck, H., Rijsselaere, T., Mateusen, B., Vyt, Ph., de Kruif, A. and Van Soom, A. (2008). AI transmitted diseases in swine: An overview. *Theriogenology* 70(8), 1337-1345.

Maes, D., Verbeke, W., Vicca, J., Verdonck, M. and de Kruif, A. (2003). Benefit to cost of vaccination against *Mycoplasma hyopneumoniae* in pig herds under Belgian market conditions from 1996 to 2000. *Livest. Sci.* 83(1), 85-93.

Maes, D. G., Deluyker, H., Verdonck, M., Castryck, F., Miry, C., Vrijens, B., Ducatelle, R. and de Kruif, A. (2001). Non-infectious herd factors associated with macroscopic and microscopic lung lesions in slaughter pigs from farrow-to-finish pig herds. *Vet. Rec.* 148(2), 41-46.

Malmsten, A., Magnusson, U., Ruiz-Fons, F., González-Barrio, D. and Dalin, A. M. (2018). A serologic survey of pathogens in wild boar (Sus scrofa) in Sweden. *J. Wildl. Dis.* 54(2), 229-237.

Marco, E., Yeske, P. and Pieters, M. (2020). Chapter 9: General control measures against Mycoplasma hyopneumoniae infections. In: Maes, D., Sibila, M. and Pieters, M. (Eds) *Book Mycoplasmas in Swine* (pp. , 163-180). Leuven, Belgium: Acco Publishers. ISBN 978-94-6379-796-2.

Marinou, K. A., Papatsiros, V. G., Gkotsopoulos, E. K., Odatzoglou, P. K. and Athanasiou, L. V. (2015). Exposure of extensively farmed wild boars (Sus scrofa scrofa) to selected pig pathogens in Greece. *Vet. Q.* 35(2), 97-101.

Martelli, F., Andres, V. M., Davies, R. and Smith, R. P. (2018). Observations on the introduction and dissemination of *Salmonella* in three previously low prevalence status pig farms in the United Kingdom. *Food Microbiol.* 71, 129-134.

Middelveld, R. J. and Alving, K. (2000). Synergistic septicemic action of the gram-positive bacterial cell wall components peptidoglycan and lipoteichoic acid in the pig *in vivo*. *Shock* 13(4), 297-306.

Muirhead, M. R. and Alexander, T. J. L. (1997). Understanding disease. In: *Managing Pig Health and the Treatment of Disease, a Reference for the Farm (Chapter 2)* (pp. 21-54). Sheffield, UK: 5M Enterprises Limited UK.

Nathues, H., Woeste, H., Doehring, S., Fahrion, A. S., Doherr, M. G. and Grosse Beilage, E. (2012). Detection of *Mycoplasma hyopneumoniae* in nasal swabs sampled from pig farmers. *Vet. Rec.* 170(24), 623.

Niemann, H. H., Schubert, W.D., and Heinz, D. W. (2004). Adhesins and invasins of pathogenic bacteria: a structural view. *Microbes Infect.* 6(1), 101-112. doi:10.1016/j.micinf.2003.11.001. PMID: 14738899.

Neirynck, W., Boyen, F., Chantziaras, I., Vandersmissen, T., Vyt, P., Haesebrouck, F., Dewulf, J. and Maes, D. (2020). Implementation and evaluation of different eradication strategies for *Brachyspira hyodysenteriae*. *Porcine Health Manag.* 6, 27.

Nicholson, T. L., Brockmeier, S. L., Loving, C. L., Register, K. B., Kehrli, M. E., Stibitz, S. E. and Shore, S. M. (2012). Phenotypic modulation of the virulent Bvg phase is not required for pathogenesis and transmission of *Bordetella bronchiseptica* in swine. *Infect. Immun.* 80(3), 1025-1036.

O'Connor, A. M., Denagamage, T., Sargeant, J. M., Rajic, A. and McKean, J. (2008). Feeding management practices and feed characteristics associated with *Salmonella* prevalence in live and slaughtered market-weight finisher swine: A systematic review and summation of evidence from 1950 to 2005. *Prev. Vet. Med.* 87(3-4), 213-228.

Office International des Epizooties (OIE). (2021). Notifiable animal diseases. Available at: https://www.oie.int/en/what-we-do/animal-health-and-welfare/animal-diseases/?_tax_animal=terrestrials%2Cmultiple-species. Accessed: 21 March 2022.

Ogwuche, A., Ekiri, A. B., Endacott, I., Maikai, B. V., Idoga, E. S., Alafiatayo, R. and Cook, A. J. (2021). Antibiotic use practices of veterinarians and para-veterinarians and the implications for antibiotic stewardship in Nigeria. *J. S. Afr. Vet. Assoc.* 92, e1-e14.

Olsen, S., Boggiatto, P. and Samartino, L. (2019). Chapter 50: Bordetellosis. In: Zimmerman, J. J., Karriker, L. A., Ramirez, A., Schwartz, K. J., Stevenson, G. W. and Zhang, J. (Eds) *Diseases of Swine* (11th edn, vol. 2019, pp. 778-791). John Wiley & Sons, Inc.

Opriessnig, T. and Coutinho, T. (2019). Chapter 53: Erysipelas. In: Zimmerman, J. J., Karriker, L. A., Ramirez, A., Schwartz, K. J., Stevenson, G. W. and Zhang, J. (Eds) *Diseases of Swine* (11th edn, vol. 2019, pp. 835-843). Hoboken, NJ: John Wiley & Sons, Inc.

Opriessnig, T., Giménez-Lirola, L. G. and Halbur, P. G. (2011). Polymicrobial respiratory disease in pigs. *Anim. Health Res. Rev.* 12(2), 133-148.

Pavlovic, M., Huber, I., Konrad, R. and Busch, U. (2013). Application of MALDI-TOF MS for the identification of food borne bacteria. *Open Microbiol. J.* 7, 135-141.

Pearson, H. E., Toribio, J. L. M. L., Lapidge, S. J. and Hernandez-Jover, M. (2016). Evaluating the risk of pathogen transmission from wild animals to domestic pigs in Australia. *Prev. Vet. Med.* 123, 39-51.

Peeters, L., Dewulf, J., Boyen, F., Brossé, C., Vandersmissen, T., Rasschaert, G., Heyndrickx, M., Cargnel, M., Pasmans, F. and Maes, D. (2019). Effects of attenuated vaccine protocols against *Salmonella Typhimurium* on *Salmonella* serology in subclinically infected pig herds. *Vet. J.* 249, 67-72.

Pieters, M. and Fano, E. (2016). *Mycoplasma hyopneumoniae* management in gilts. *Vet. Rec.* 178(5), 122-123.

Pieters, M. and Maes, D. (2019). Chapter 56: Mycoplasmosis. In: Zimmerman, J. J., Karriker, L. A., Ramirez, A., Schwartz, K. J., Stevenson, G. W. and Zhang, J. (Eds) *Diseases of Swine* (11th edn, vol. 2019, pp. 863-883). Hoboken, NJ: John Wiley & Sons, Inc.

Post, K. (2019). Chapter 47: Overview of bacteria. In: Zimmerman, J. J., Karriker, L. A., Ramirez, A., Schwartz, K. J., Stevenson, G. W. and Zhang, J. (Eds) *Diseases of Swine* (11th edn, vol. 2019, pp. 745-749). Hoboken, NJ: John Wiley & Sons, Inc.

Prieto-Martinez, N., Bussalleu, E., Garcia-Bonavila, E., Bonet, S. and Yeste, M. (2014). Effects of Enterobacter cloacae on boar sperm quality during liquid storage at 17 degrees C. *Anim. Reprod. Sci.* 148(1-2), 72-82.

Ramirez, A. (2018). Chapter 1: Diseases affecting pigs: An overview of common bacterial, viral and parasitic pathogens of pigs. In: Wiseman, J. (Ed.) *Achieving Sustainable Production of Pig Meat* (vol. 3, p. 328). Burleigh Dodds Science Publishing: Animal Health and Welfare.

Randall, L. P., Lemma, F., Koylass, M., Rogers, J., Ayling, R. D., Worth, D., Klita, M., Steventon, A., Line, K., Wragg, P., Muchowski, J., Kostrzewa, M. and Whatmore, A. M. (2015). Evaluation of MALDI-ToF as a method for the identification of bacteria in the veterinary diagnostic laboratory. *Res. Vet. Sci.* 101, 42-49.

Register, K. and Brockmeier, S. (2019). Chapter 57: Pasteurellosis. In: Zimmerman, J. J., Karriker, L. A., Ramirez, A., Schwartz, K. J., Stevenson, G. W. and Zhang, J. (Eds)

Diseases of Swine (11th edn, vol. 2019, pp. 884–897). Hoboken, NJ: John Wiley & Sons, Inc.

Regula, G., Lichtensteiger, C. A., Mateus-Pinilla, N. E., Scherba, G., Miller, G. Y. and Weigel, R. M. (2000). Comparison of serologic testing and slaughter evaluation for assessing the effects of subclinical infection on growth in pigs. *J. Am. Vet. Med. Assoc.* 217(6), 888–895.

Rhouma, M., Fairbrother, J. M., Beaudry, F. and Letellier, A. (2017). Post weaning diarrhea in pigs: Risk factors and non-colistin-based control strategies. *Acta Vet. Scand.* 59(1), 31.

Rycroft, A. (2020). Chapter 1: Overview of the general characteristics and classification of porcine *Mycoplasma* species. In: Maes, D., Sibila, M. and Pieters, M. (Eds) *Book Mycoplasmas in Swine* (pp. 25–46). Leuven, Belgium: Acco Publishers,. ISBN 978-94-6379-796-2.

Schulze, M., Ammon, C., Rudiger, K., Jung, M. and Grobbel, M. (2015). Analysis of hygienic critical control points in boar semen production. *Theriogenology* 83(3), 430–437.

Sepúlveda, L., Bussalleu, E., Yeste, M. and Bonet, S. (2014). Effects of different concentrations of *Pseudomonas aeruginosa* on boar sperm quality. *Anim. Reprod. Sci.* 150(3–4), 96–106.

Sepúlveda, L., Bussalleu, E., Yeste, M. and Bonet, S. (2016). Effect of *Pseudomonas aeruginosa* on sperm capacitation and protein phosphorylation of boar spermatozoa. *Theriogenology* 85(8), 1421–1431.

Sepúlveda, L., Bussalleu, E., Yeste, M., Torner, E. and Bonet, S. (2013). How do different concentrations of *Clostridium perfringens* affect the quality of extended boar spermatozoa? *Anim. Reprod. Sci.* 140(1–2), 83–91.

Songer, J. G. and Post, K. W. (2005). Origin and evolution of virulence. In: *Veterinary Microbiology: Bacterial and Fungal Agents of Animal Disease* (pp. 3–9). St. Louis, MO: Elsevier Saunders.

Tack, D., Ray, L., Griffin, P., Cieslak, P., Dunn, J., Rissman, T., Jervis, R., Lathrop, S., Muse, A., Duwell, M., Smith, K., Tobin-D'Angelo, M., Vugia, D., Zablotsky-Kufel,J., Wolpert, B., Tauxe, R. and Payne,D. (2020). Preliminary incidence and trends of infections with pathogens transmitted commonly through food – Foodborne diseasesactive surveillance network. *MMWR Morb. Mortal. Wkly Rep. (MMWR)* 1069(17), 509–514.

Thomas, S. R. and Elkinton, J. S. (2004). Pathogenicity and virulence. *J. Invertebr. Pathol.* 85(3), 146–151.

Thomson, J. and Friendship, R. (2019). Chapter 15: Digestive system. In: Zimmerman, J. J., Karriker, L. A., Ramirez, A., Schwartz, K. J., Stevenson, G. W. and Zhang, J. (Eds) *Diseases of Swine* (11th edn, vol. 2019, pp. 234–263). Hoboken, NJ: John Wiley & Sons, Inc.

Tobias, T. J., Bouma, A., van den Broek, J., van Nes, A., Daemen, A. J. J. M., Wagenaar, J. A., Stegeman, J. A. and Klinkenberg, D. (2014). Transmission of *Actinobacillus pleuropneumoniae* among weaned piglets on endemically infected farms. *Prev. Vet. Med.* 117(1), 207–214.

Úbeda, J. L., Ausejo, R., Dahmani, Y., Falceto, M. V., Usan, A., Malo,C. and Perez-Martinez, F. C. (2013). Adverse effects of members of the Enterobacteriaceae family on boar sperm quality. *Theriogenology* 80(6), 565–570.

Uzal, F. and Songer, G. (2019). Chapter 51: Clostridial diseases. In: Zimmerman, J. J., Karriker, L. A., Ramirez, A., Schwartz, K. J., Stevenson, G. W. and Zhang, J. (Eds) *Diseases of Swine* (11th edn, vol. 2019, pp. 792–806). John Wiley & Sons, Inc.

Vannucci, F., Gebhart, C. and McOrist, S. (2019). Chapter 58: Proliferative enteropathy. In: Zimmerman, J. J., Karriker, L. A., Ramirez, A., Schwartz, K. J., Stevenson, G. W. and Zhang, J. (Eds) *Diseases of Swine* (11th edn, vol. 2019, pp. 898-911). John Wiley & Sons, Inc.

Vanrolleghem, W., Tanghe, S., Verstringe, S., Bruggeman, G., Papadopoulos, D., Trevisi, P., Zentek, J., Sarrazin, S. and Dewulf, J. (2019). Potential dietary feed additives with antibacterial effects and their impact on performance of weaned piglets: A meta-analysis. *Vet. J.* 249, 24-32.

Wu, N., Abril, C., Thomann, A., Grosclaude, E., Doherr, M. G., Boujon, P. and Ryser-Degiorgis, M. P. (2012). Risk factors for contact between wild boar and outdoor pigs in Switzerland and investigations into potential *Brucella suis* spill-over. *BMC Vet. Res.* 8, 116.

Zeeh, F., Vidondo, B. and Nathues, H. (2020). Risk factors for the infection with *Brachyspira hyodysenteriae* in pig herds. *Prev. Vet. Med.* 174, 104819.

Chapter 3

Improving gut function in pigs to prevent dysbiosis and post-weaning diarrhoea

Charlotte Lauridsen, Ole Højberg and Nuria Canibe, Aarhus University, Denmark

1 Introduction
2 Eubiosis versus dysbiosis
3 Dysbiosis and post-weaning diarrhoea (PWD)
4 Microbiota composition as a predictor of post-weaning diarrhoea risk
5 Nutritional and dietary strategies to prevent dysbiosis in relation to post-weaning diarrhoea
6 Host factors influencing gut function
7 Conclusion and future trends in research
8 Where to look for further information
9 References

1 Introduction

Post-weaning diarrhoea (PWD) is a significant enteric disease resulting in considerable economic loss for the pig industry. The development of PWD in pigs, which requires treatment with antibiotics, can be attributed to many factors. During the first 2-3 weeks post-weaning, diarrhoea is mostly associated with *Escherichia coli* infections of the small intestine, which may be initiated by enteric dysbiosis. However, when the pigs are older and heavier, diarrhoea may be caused by a combination of various pathogenic bacteria (*E. coli*, *Lawsonia*, *Brachyspira*) and undigested nutrients, and is associated with dysbiosis in the colon. The term 'dysbiosis' refers to an ecosystem where bacteria no longer live together in mutual harmony. The development of PWD can be considered in a simplistic manner, that is, dysbiosis arises when the commensal bacteria no longer control the potential pathogenic bacteria, as diarrhoea is induced when pathogenic bacteria colonise and adhere to the gut epithelium. There are a number of factors that can improve gut function to enhance the epithelial barrier and immune system, and thus prevent PWD, with the prevention of dysbiosis playing a major role. The overall objective of this chapter is to provide

http://dx.doi.org/10.19103/AS.2021.0089.15

an overview of factors that may enhance gut function in terms of a balanced or eubiotic ecosystem and epithelial barrier function.

2 Eubiosis versus dysbiosis

The eubiotic microbiota is defined as a health-promoting microbiota. A gut microbiota in a eubiotic state metabolises indigestible compounds, supplies essential nutrients, defends against colonisation by opportunistic pathogens, and contributes to the development of the intestine and immune system (Round and Mazmanian, 2009, Wang et al., 2017). Further, the importance of maintaining the gut microbiota in a eubiotic state is becoming increasingly evident as studies indicate interrelations between the gut microbiota and other organs of the host, for example, the gut-brain axis and the gut-lung axis (Kraimi et al., 2019, Nair et al., 2019, Wang et al., 2020).

A wide range of ecological interactions occur within microbial communities: (i) *mutualism*, where both participants benefit; (ii) *amensalism*, where one partner is harmed without any advantage to the other; (iii) *commensalism*, where one partner obtains an advantage without any help or harm to the other; (iv) *competition*, where both participants harm each other; and (v) *predation and parasitism*, where one benefits while the other is harmed (Faust and Raes, 2012, Costa et al., 2014). This implies that the presence or abundance of one member of the microbiota can have an impact on the presence or abundance of others, and depending on the role of the involved microorganisms, will affect the population composition and its functionality to differing extents.

The nature of these interactions, in addition to known and unknown factors related to the host, for example, genotype, age, gender etc., and environment, pH, oxygen level, substrate availability, digestive secretions and transit time, shape the succession and stability of the gut microbiota.

All this simply means that the intestinal microbiome is composed of multiple microbial species in competition for limited nutrients and attachment sites, and with varying susceptibility to internal and external conditions and perturbations. However, and unfortunately, although simple to summarise, the manner in which the microbiota composition and diversity are shaped and persist is still poorly understood (Mosca et al., 2016). In this context, there is some evidence from humans showing that succession is not a passive process but is actively guided by the host. For example, it has been shown that the infant gut actively recruits *Bifidobacteria* and *Bacteroides thetaiotaomicron* via the secretion of fucosylated oligosaccharides (FOS) into the gut lumen (Mao and Franke, 2015). In addition, the ability to metabolise endogenous glycans from host mucus (O-linked glycans) is likely to be a key factor in determining which microorganisms physically associate with this layer (Koropatkin et al., 2012). Hence, the type of endogenous glycans from host mucus glycans contributes to determining the composition of the gut microbiota. On the other hand, the initial colonising bacteria can, in turn, modulate host gene expression, influencing the successive microbial microbiota (Hooper et al., 2001). The presence of commensal bacteria has a direct influence on immune maturation, and microbial exposure influences the expression of a large number of immune-related genes (Mulder et al., 2009). High inter-individual variation in such phenomena and specific characteristics would help explain the high diversity of gut microbiota composition among individuals.

It is widely recognised that the gut microbiota has a huge impact on the health of the host (Chowdhury et al., 2007, Parker et al., 2018), and more recently, data have shown that the gut microbiota can also influence growth performance (Gardiner et al., submitted) and host behavior (Brunberg et al., 2016, Sylvia and Demas, 2018). Consequently, maintenance of a balanced gut microbiota is widely considered to be essential for intestinal homeostasis and host health, performance and welfare (Zeng et al., 2017, Patil et al., 2020). A balanced microbiota does not mean a static microbiota. Although there seems to be a core microbiota found in most pigs (Kim et al., 2011a, Holman et al., 2017), longitudinal studies have shown that the microbiota composition of the pig gut changes with age (Kim et al., 2011a, Frese et al., 2015, Mach et al., 2015, Slifierz et al., 2015, Bian et al., 2016, Ke et al., 2019). In these studies, some changes are related to dietary practices or other factors that depend on the animal's age during the pig production process, but alterations in microbiota composition due to age itself have also been observed.

In contrast to 'eubiosis,' 'dysbiosis' is broadly defined as any change in the composition of resident commensal microbiota relative to the microbiota found in healthy individuals (Petersen and Round, 2014). Given the known importance of the microbiota on host health, development performance and welfare, it has been speculated that these observed changes in microbial composition are contributing factors to the initiation and/or persistence of disease (Petersen and Round, 2014). The effort in understanding dysbiosis is driven by the hypothesised relationship between dysbiosis and increased risk of disease. To better understand this relationship, a more complete definition of dysbiosis, which mentions the consequences that changes in the microbiota community have on functionality, is 'gut dysbiosis refers to altered composition of the gut microbiota that is associated with functional changes in the microbial transcriptome, proteome or metabolome' (Zeng et al., 2017). For further discussion on the definition of dysbiosis, see (Hooks and O'Malley, 2017).

However, when strictly defined according to microbiota composition, dysbiosis can be categorised into three types: (i) loss of beneficial microbial organisms; (ii) expansion of pathobionts or potentially harmful microorganisms; and (iii) loss of overall microbial diversity. It is likely that dysbiosis includes all three of these manifestations concomitantly to influence the outcome (Petersen and Round, 2014).

There is a connection between low diversity and dysbiosis in macro-ecosystems, which can also explain processes in the gastrointestinal ecosystem of pigs (Lozupone et al., 2012). For example, a high biodiversity is thought to help maintain a stable ecosystem function, that is, it provides robustness/stability/resilience. This occurs through the versatile responses of such a system to external or internal changes (Mosca et al., 2016). On the contrary, low diversity is associated with impaired resilience, that is, a diverse ecosystem would be able to cope with a disturbance which a less diverse ecosystem would not, leading to dysbiosis in the latter but not the former. The loss of microbial diversity characterises a dysbiotic state, and low microbial diversity is believed to increase the risk of dysbiosis when the system is subjected to perturbation (Mosca et al., 2016).

In humans, the loss of microbiota diversity is common in intestinal and extra-intestinal disorders (Mosca et al., 2016). However, it should be kept in mind that the contexts in which these relationships are observed are often chronic diseases/disorders such as allergy, inflammatory bowel diseases, diabetes mellitus, multiple sclerosis, colorectal cancer, obesity etc. Factors such as a Western lifestyle, leading to loss of microbial diversity, are suspected to result in a less robust system and thereby increase the risk of disease (Mosca et al., 2016). When considering pigs, the types of diseases of concern when focusing on the gut microbiota are most often non-chronic digestive diseases, PWD being of paramount importance. Post-weaning diarrhoea is not a chronic disease and is a result of multiple factors associated with abrupt changes in diet as well as social and physical environment that piglets are subjected to at an age when their digestive and immune system are not sufficiently developed (Pluske, 2016, Gresse et al., 2017). In addition, antibiotic use, supplemented in the feed or as therapeutic treatment, commonly occurs during this phase and is known to cause loss of biodiversity in the gut microbiota (Gresse et al., 2017).

Whether the relationship between dysbiosis and disease can be considered as a correlation or causality is a subject of debate; that is, is dysbiosis the consequence or cause of the disease? One specific case illustrating how dysbiosis leads to a non-chronic disease (which would resemble the situation of PWD in piglets) is diarrhoea caused by the blooming of *Clostridium difficile*, typically observed in humans as a result of antibiotic treatment. *C. difficile*-associated diarrhoea is a serious condition with high mortality, especially in frail elderly people. The disruption of the microbiota (dysbiosis) due to antibiotic treatment, often seen as a decrease of microbiota diversity, is considered the main driver of the disease (Chang et al., 2008, Britton and Young, 2012, Bien et al., 2013, Lewis et al., 2015). Niches occupied by commensals in healthy individuals become available to *C. difficile*, which can then overgrow and cause disease (Chang et al., 2008). In ecological terms, it could be said that typical K-strategists (dominating stable/climax ecosystems) are reduced, allowing the temporary bloom of r-strategists (early colonisers).

However, sometimes it is not dysbiosis that favours pathogen overgrowth, but pathogen proliferation that causes dysbiosis, and consequently, disease development. Direct competition of the pathogen/s with the commensal microbiota or indirect competition by promoting specific responses in the host, facilitates growth of the pathogen. For example, alterations in oxygen concentration, at the expense of members of the commensal microbiota, could lead to dysbiosis and disease (Gresse et al., 2017, Zeng et al., 2017). It has been shown that *Salmonella* can alter the normal composition of the gut microbiota in mice, which is associated with *Salmonella* virulence factors inducing inflammatory mucosal host responses that favour pathogen growth (Barman et al., 2008). Arguello et al. (Arguello et al., 2018) described how infection with *Salmonella typhimurium* caused dysbiosis in piglets, with the host's response to infection (immune response and metabolic changes) probably being the main cause of disruption to the microbial composition post-infection (Guevarra et al., 2019). Similar conclusions were drawn by (Drumo et al., 2016), who found that *S. typhimurium* induced inflammation and reduced specific protective microbiota species (short-chain fatty acid (SCFA)-producing bacteria) normally involved in providing a barrier against pathogens.

There are also examples in which the overgrowth of pathogens is specifically stimulated, leading to dysbiosis and disease. Although the mechanisms behind the impact of dietary protein on PWD are not fully understood, a contributing factor is thought to be an excess of dietary fermentable protein, which stimulates the growth of proteolytic pathogens such as *E. coli* or *Clostridium perfringens*, due to increased substrate availability and higher pH (Nyachoti et al., 2006, Pluske, 2013, Pieper et al., 2016), resulting in dysbiosis and disease, namely, diarrhoea. Undigested dietary protein is further converted into toxic metabolites which can impair epithelial integrity and promote inflammatory reactions, as described below.

3 Dysbiosis and post-weaning diarrhoea (PWD)

Piglets are separated from their mothers at weaning, which itself triggers a stressful reaction, leading to an abrupt change in the diet – from a milk-based diet to a diet containing plant-based substrates. Animals at this age have an immature digestive system, with the depletion of absorbed nutrients occurring concomitantly with a high substrate availability to the gastrointestinal microbiota, leading to a high risk of unbalancing the community composition. This abrupt change in diet also often results in a period of anorexia or highly reduced feed intake (Bruininx et al., 2002), which may lead to undernutrition of the host as well as shortage of nutrients to the microbiota. Further, the loss of immune protection from the mother's milk at a stage where the immune system is not fully developed (Stokes et al., 2004) leaves the animals more vulnerable to infections from pathogenic enteric bacteria. Moreover, the piglets are exposed to various stress factors such as transportation and mixing with unfamiliar animals, which further increases the risk of gastrointestinal microbial alterations, that is, dysbiosis and infections (Gresse et al., 2017, Pluske, 2016) (Fig. 1). As reviewed elsewhere (Gresse et al., 2017), this weaning transition results in a number of changes in the piglet's gastrointestinal tract, including activation of pathways related to inflammatory responses, changes in hormonal activity, reduction in gastric motility, induction of small intestine atrophy and reduced villous height, reduction in nutrient, fluid and electrolyte absorption and increased permeability to antigens and toxins. This illustrates the difficulty of identifying a single parameter as the cause of PWD, as it is a result of multifactorial events. It is now increasingly recognised that imbalance or disruption of the gut microbiota at weaning, or dysbiosis, is one of the key factors in the aetiology of PWD (Gresse et al., 2017).

In nature, weaning occurs at a later age and much more gradually, compared with commercial pig production, and these factors (i.e. later age and gradual weaning) are considered to be key elements in reducing the risk of developing PWD in piglets (Madec et al., 1998, Wellock et al., 2007, Pluske, 2016). However, the current practice of conventional modern intensive pig

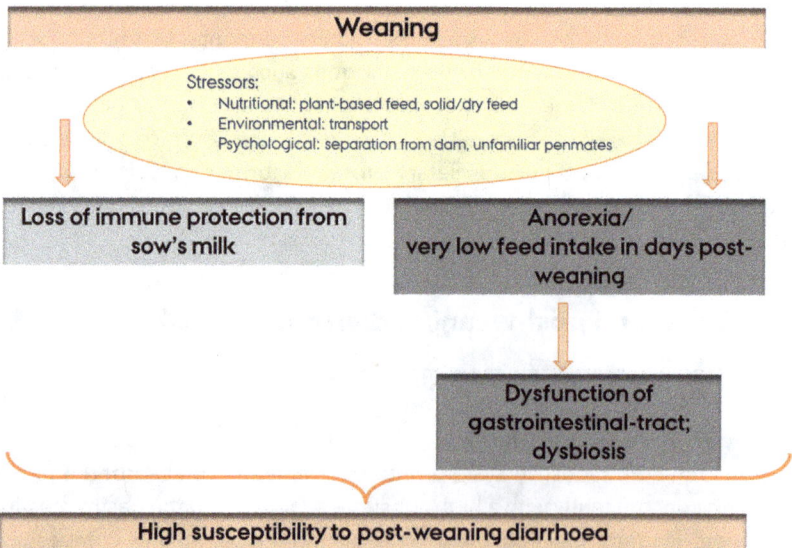

Figure 1 Factors inducing dysbiosis around weaning enhancing the risk of enteric diseases.

production makes it difficult to wean piglets at a later age, whereas introducing feeding interventions while suckling the sow may offer the possibility of adapting the enteric system and microbiota more gradually to the weaning situation. If we assume that the pig's metabolism, with all its complex systems and organs including the digestive system, is adapted to the dynamics occurring in nature, the modern pig production practice of weaning interferes with an otherwise well-functioning system at a time when it is unprepared for this disturbance. This is probably the overall key aspect leading to PWD. It is not only a matter of preventing gut microbiota dysbiosis but also of improving the animal's robustness to cope with the changes occurring at weaning. Since diet is one of the factors clearly affecting the composition of the gastrointestinal tract microbiota, changing from the sow's milk to a plant-based commercial pig diet will affect the gastrointestinal tract microbiota. Therefore, even if all the other mentioned factors occurring at weaning were unchanged, the GI tract microbiota would be affected by weaning. However, the question remains, when do we consider changes in microbiota composition as a logical and 'normal' consequence of changes in substrate availability and when do we consider it to be dysbiosis and thereby, a situation with increased risk of PWD?

The results of some studies investigating the impact of weaning on faecal microbiota composition are summarised in Table 1, showing microbial data (mostly at the genus level) registered in the period around weaning, approximately two weeks pre- and post-weaning. As expected, there is no consensus in the results obtained; however, some general trends can be observed. The most consistent pattern is the decrease in *Bacteroides* abundance and increase in *Prevotella* abundance observed post-weaning compared with the pre-weaning microbiota (Pajarillo et al., 2014, Mach et al., 2015, Chen et al., 2017, Guevarra et al., 2018, Luise et al., 2021, Massacci et al., 2020). The higher abundance of *Bacteroides* during the pre-weaning period is suggested to be due to the utilisation of mono- and oligosaccharides present in sow's milk, whereas the increased abundance of *Prevotella* during the post-weaning period is due to the ability to degrade polysaccharides of plant origin (Pajarillo et al., 2014, Mach et al., 2015, Guevarra et al., 2018). Another often reported observation is a high abundance of *Fusobacterium*/Fusobacteriaceae pre-weaning, with a decrease post-weaning (Bian et al., 2016, Poulsen et al., 2018b, Ke et al., 2019). Fusobacteria play a role in a wide spectrum of human infections (Bennett and Eley, 1993). The abundance of bacteria belonging to the phylum Fusobacteria, or the family Fusobacteriaceae, was increased in piglets suffering from neonatal porcine diarrhoea (Hermann-Bank et al., 2015, Yang et al., 2017), and an increased abundance of *Fusobacterium* has also been observed in diarrhoeic piglets infected with the porcine epidemic diarrhoea virus. There are also species within the genus *Fusobacterium* that inhibit the growth of pathogenic bacteria by producing bacteriocin-like compounds

Table 1 Microbiota diversity and the most abundant, or changes in abundance, of microbiota members in faeces during the pre- and immediate post-weaning period (up to 14 d post-weaning)

Weaning day and sampling time points	Diversity	Pre-weaning	Post-weaning	Ref
Weaning d28 d28 of age; d14 post-weaning	Somewhat higher diversity (Shannon) post-weaning	Highest pre-weaning:*Bacteroides* *Dorea* *Escherichia* *Fusobacterium* *Blautia* *Campylobacter* *Phascolarctobacterium* *Parabacteroides* *Subdulogranulum* *Victivalis* *Ruminococcus* *Suturella*	Highest post-weaning:*Prevotella* *Clostridium*	1
Weaning d21 d1, 3, 5, 7, 14 and 21 of age; d7 and 14 post-weaning	Diversity similar pre- and post-weaning	Highest pre-weaning:Enterobacteriaceae Bacteroidaceae Clostridiaceae Enterococcaceae Fusobacteriaceae Porphyromonadaceae Verrucomicrobiaceae Rikenellaceae Peptostreptococcaceae	Highest post-weaning:Prevotellaceae Ruminococcaceae Lactobacillaceae Veillonellaceae Streptococcaceae Succinivibrionaceae Spirochaetaceae	2

Weaning d28 d14 of age; d8 post-weaning		Higher pre-weaning:Bacteroides Butyricimonas Oscillibacter Clostridium sensu stricto Clostridium IV Clostridium XIVa Escherichia/Shigella	Higher post-weaning:Prevotella Acetivibrio Oribacterium Paraprevotella Roseburia Succinivibrio	3
Weaning d21 d7-21 of age; d7-14 post-weaning		Most abundant:Clostridium Escherichia Lactobacillus Clostridium XIVa Unclassified Firmicutes	Most abundant:Megasphaera Lactobacillus Clostridium Unclassified Firmicutes Succinivibrio	4
Weaning day; d10 post-weaningDiversity increased		Higher pre-weaning:Megasphaera Escherichia/ Shigella Bacteroides Fusobacterium Phascolarctobacterium	Higher post-weaning:Blautia Paraprevotella Roseburia Clostridium sensu stricto Prevotella	5
Weaning d21 d21 of age; d3 post-weaning (colon samples)	Diversity similar; richness lower post-weaning	Higher pre-weaning:Porphyromonadaceae Alloprevotella Barnesiella Oscillibacter	Higher post-weaning:Lachnospiraceae Negativicutes Selenomonadales Campylobacterales	6
Weaning d28d28 of age; d14 post-weaning	Diversity increased post-weaning	Most abundant:Prevotella Fusobacterium Bacteroides Oscillopira Lactobacillus Ruminococcus	Most abundant:Prevotella Lactobacillus Oscillopira Ruminococcus Coprococcus Streptococcus	7

(Continued)

Table 1 (Continued)

Weaning day and sampling time points	Diversity	Pre-weaning	Post-weaning	Ref
Weaning d26d26 of age; d9 post-weaning	Diversity increased post-weaning	Highest pre-weaning:Bacteroides Christensenellaceae Enterobacteriaceae Clostridiaceae Oscillospira Ruminococcus Porphyromonadaceae Fusobacteriaceae Enterococcaceae Rikenellaceae	Highest post-weaning:Prevotella Blautia Lachnospira Roseburia Faecalibacterium Succinivibrionaceae Veillonellaceae Peptostreptococcaceae	8
Weaning d28d14 of age; d14 post-weaning	Increased alpha diversity with age	Highest pre-weaning:Bacteroidaceae Enterobacteriaceae Fusobacteriaceae Enterococcaceae Odoribacteraceae	Highest post-weaning:Prevotellaceae Veillonellaceae Paraprevotellaceae S24-7 Peptostreptococcaceae Spirochaetaceae Ruminococcaceae	9
Weaning d14, 21, 28 and 42 Weaning day; d7 post-weaning	Diversity and richness increases, except weaning d42	Highest pre-weaning:Bacteroides Ruminococcus Oscillospira Clostridium	Highest post-weaning:Succinivibrio Prevotella Campylobacter	10

Source: 1. Pajarillo et al., 2014; 2. Frese et al., 2015; 3. Mach et al., 2015; 4. Slifierz et al., 2015; 5. Chen et al., 2017; 6. Li et al., 2018b; 7. Poulsen et al., 2018b; 8. Luise et al., 2021; 9. Motta et al., 2019; 10(Massacci et al., 2020).

(Portrait et al., 2000); therefore, the impact of Fusobacteria in general on suckling pigs remains unclear.

Using denaturing gradient gel electrophoresis to characterise changes in the *Lactobacillus* population in the ileal digesta around weaning, Janczyk et al.(Janczyk et al., 2007) observed that the *Lactobacillus* population undergoes dramatic, partly reversible changes as a consequence of weaning. The diversity of lactobacilli was high at weaning and remained unchanged for the first two days post-weaning, it then decreased five days after weaning, and was restored to initial levels eleven days after weaning. The abundance of *L. sobrius* decreased one day post-weaning, followed by an increase eleven days post-weaning, which is in line with studies showing *L. sobrius* as a dominant species in the ileal digesta of pigs. At 1–2 days post-weaning, the lactobacilli profiles were similar to those at weaning, whereas at 11 days post-weaning, the profiles differed greatly from those at weaning. Some changes in the *Lactobacillus* population profile were explained by dietary changes, such as a reduction of *L. johnsonii* post-weaning, which is known for its proteolytic activity, that is, casein degradation in milk. However, this is in contrast with the findings of (Pajarillo et al., 2014), who reported an increased abundance of *L. johnsonii* 2 weeks post-weaning and the restoration of *L. sobrius*, known to ferment starch, 11 days post-weaning. These are just a few examples of changes in microbiota composition that occur post-weaning, which can be explained by the introduction of a new diet. These changes are inevitable and necessary in order to help the host extract energy from the plant-based ingested substrates.

Although the faecal microbiota changes throughout the life of the pig (Kim et al., 2011a, Slifierz et al., 2015, Ke et al., 2019), a relative degree of stability occurs at around three weeks post-weaning (Slifierz et al., 2015, Bian et al., 2016). Slifierz et al. (2015) sampled piglets on days 1, 3, 7, 14, 21 (prior to weaning), 28, 35, 42 and 49, and although weaning played an important role, ageing appeared to be the most significant driver of faecal microbiota development.

These results illustrate that changes in microbiota composition around weaning are inevitable, although they can, to some extent, be controlled by specific and more gradual changes in diet composition and less abrupt management practices around weaning. Upon considering these results, a distinction should be made between changes in microbiota composition due to the 'natural' dynamics occurring in relation to weaning and a dysbiotic microbiota *per se*.

According to the definition of dysbiosis, which is any change in the composition of resident commensal communities relative to the community found in healthy individuals (Petersen and Round, 2014), the changes in microbiota as a consequence of weaning would not be classified as a dysbiotic microbiota, as the animals do not suffer from disease. However, they could be more prone to disease if challenged or exposed to a pathogenic burden at that stage, which is the extended hypothesis regarding the consequence/risk of the changes in microbiota composition occurring after weaning.

During the first two weeks after weaning, diarrhoea is most often caused by enterotoxigenic *E. coli* (ETEC); therefore, a key question is, what favours the proliferation and colonisation of ETEC in the gut of piglets during the immediate post-weaning period?

Studies investigating the pathology of inflammatory bowel disease (IBD) in humans indicate that the bloom of *E. coli* appears to be a consequence rather than a cause of inflammation in IBD, likely due to the unique ability of *E. coli* to thrive in the inflamed gut (Zeng et al., 2017).

Inflammation in the piglet gut can be triggered by the weaning process itself, with all the abrupt changes that the animals are subjected to, such as antibiotic treatment, injury, anorexia, etc. (Gresse et al., 2017). Some of the characteristics of the inflamed gut that give *E. coli* an advantage to bloom are (Gresse et al., 2017, Zeng et al., 2017):

- The amount of nitrate in the inflamed intestine rises rapidly. This nitrate-rich tissue environment confers a growth advantage for Enterobacteriaceae, such as *E. coli*, through nitrate respiration, as genes encoding nitrate reductases are found within the genomes of facultative anaerobic Enterobacteriaceae but are largely absent in obligate anaerobic bacteria.
- The higher blood flow and haemoglobulin within the luminal environment in an inflamed intestine is thought to result in a more aerobic microenvironment in the lumen that favours the bloom of facultative anaerobes such as Enterobacteriaceae.
- As a response to inflammation, an increased production of mucins occurs in the intestine, probably to impair the colonisation of pathogens. Sialic acid is one of the major carbohydrates in mucins, released after sialidase activity. It has been shown that sialic acid release from host glycans provided an intestinal niche for *E. coli* growth in mice with colitis. Sialic acids are taken up by bacteria like Enterobacteriaceae, lacking *de novo* biosynthetic pathways for these amino sugar moieties, and incorporated into bacterial capsules and lipooligosaccharides. This protects microbes from recognition by the host immune system and affects the host immune response via interactions (Huang et al., 2015). In addition, sialic acids are a prime target for bacterial adhesins and toxins from various pathogens, including *E. coli*, which is important for the pathogenic process (Huang et al., 2015).

Further, mucin-degrading bacteria proliferate during intestinal inflammation and mediate the release of less complex sugars from mucins. The accumulation of these compounds may either directly or indirectly lead to the depletion of commensal bacteria in the inflamed gut and confer a growth advantage to Enterobacteriaceae as well as pathogens such as *S. enterica* serovar Typhimurium and *C. difficile* (Stecher, 2015, Zeng et al., 2017).

- Members of Enterobacteriaceae harbour a high number of siderophores, which are low-molecular weight compounds that chelate ferric iron with high affinity. The complex is actively transported across the membranes of Gram-negative bacteria, and siderophores are therefore crucial for their survival in low-iron environments. In the inflamed gut, a variety of antimicrobial proteins are produced and released by host cells to limit the availability of iron for bacterial growth. A study by Singh et al. (2015) showed that the siderophore enterobactin, released by *E. coli*, inhibits the activity of the neutrophil bactericidal enzyme myeloperoxidase, hence promoting the survival of *E. coli* in the inflamed gut.

All these factors contribute to explain how inflammatory reactions of the gut epithelium, occurring in the post-weaning period due to external factors such as antibiotic use, dietary changes and stressors, lead to a number of changes in the gut environment that favour the blooming of *E. coli* via inflammation-driven fitness factors to overcome environmental and nutritional challenges (Zeng et al., 2017).

4 Microbiota composition as a predictor of post-weaning diarrhoea risk

What makes a piglet more prone to PWD? Dysbiosis due to low feed intake, stress, antibiotics etc. at weaning, which increase the risk of diarrhoea, and the characteristics of the microbiota during the suckling period and before diarrhoea occurs can result in predisposition to PWD. In order to reduce PWD, manipulation of the factors determining PWD is crucial. In this respect, the role of early-life structure and function of the gut microbial community and their function in pathogenesis of PWD are of great interest. However, much remains unknown.

Perhaps piglets are weaned at a developmental stage/age in which the microbial ecosystem is not in a dysbiotic state but cannot cope with the disturbances that it is subjected to. This means that we are interfering with an otherwise well-functioning system at the wrong moment. Evolutionarily, animals are not 'programmed' to possess the complete arsenal of defence mechanisms against all possible threats at all times, but only at times where each threat naturally appears. If we intend to interfere with this timing, we should first make the enteric system robust or resilient for the disturbance applied to the microbiota or host.

Investigations trying to establish a causal link between microbiota composition during the eubiotic state and the later diarrhoeic state are important, and would help define the aim of strategies to prevent PWD.

Studies have shown differences in the microbiota composition of healthy animals that later exhibited differing degrees of infection, indicating that the microbiota composition might predispose or protect against disease development, that is, pathogen colonisation. For example, Bearson et al. (2013) observed significant differences between the faecal microbiota of high-shedder and low-shedder pigs before inoculation with *S. enterica* serovar Typhimurium. The higher abundance of the Ruminococcaceae family in the low-shedders compared with high-shedders was reported, but no differences in diversity indices, such as richness and evenness, were measured.

Dou et al. (2017) observed lower evenness, higher diversity and higher abundance of the Lachnospiraceae, Ruminococcaceae and Prevotellaceae families and lower abundance of the Fusobacteriaceae and Corynebacteriaceae families in seven-day-old piglets that did not develop PWD compared with those that developed PWD. *Roseburia*, *Prevotella* and genera belonging to Ruminococcaceae may provide benefits to the host, since species from these genera are adapted to metabolise a wide range of complex oligo- and polysaccharides. The higher abundance of the Lachnospiraceae, Ruminococcaceae and Prevotellaceae families in healthy pigs may provide a higher energy-harvesting capacity in suckling piglets that do not have PWD and an adequate prevention strategy against pathogen infection (Dou et al., 2017).

On the other hand, Costa et al. (2014) observed no significant differences in the pre-inoculation microbiomes of pigs regarding diversity (Chao and Shannon indices) or taxonomic composition that did or did not develop mucohaemorrhagic diarrhoea, following challenge with *Brachyspira hampsonii*.

Data of this type, showing microbiota composition as a predictor for risk of PWD development, are important but scarce, and therefore, conclusions on specific early-life microbiota profiles to be targeted to reduce the risk of PWD are difficult to determine. It should also be kept in mind that it is the microbiota function, rather than composition, that is relevant for the impact on the host, but this aspect is even less elucidated.

5 Nutritional and dietary strategies to prevent dysbiosis in relation to post-weaning diarrhoea

As stated above, a eubiotic gut microbiota benefits the host by conferring protection against pathogen colonisation, shaping the immune response development and promoting gut development etc. Consequently, the search for strategies to avoid alteration of the microbiota profile from a eubiotic to a dysbiotic state has been ongoing for many decades.

As ETEC are the main pathogens causing PWD, as stated by (Zeng et al., 2017), understanding the factors in the inflamed gut that confer a growth advantage to Enterobacteriaceae will shed light on ways to prevent the overgrowth of enterobacterial species in the gut from altering host homeostasis and contributing to disease development. This is an important area of research as increased knowledge is essential, with the optimal solution obviously being to prevent inflammation of the gut during the post-weaning period. Since this inflammation is the result of various factors, including reduced feed intake, use of in-feed antibiotics, an abrupt change to solid and plant-based feed, stress etc. (Gresse et al., 2017), there is a general consensus that one single tool would most probably be unsuccessful, and therefore, several initiatives with different modes of action are needed to successfully prevent PWD.

Strategies involving management, genetic and nutritional approaches have been investigated. Non-dietary strategies include moving the sow and leaving the piglets in the farrowing pen, increasing the weaning age, intermittent suckling, multisuckling and breeding programs for selection of pigs resistance to F4 ETEC infection (Pluske, 2016, Turpin et al., 2017). These strategies aim to reduce stress, promote immune maturation or avoid pathogen attachment, and thereby increase robustness of the animals and prevent pathogen proliferation and/or dysbiosis. The nutritional approaches will now be discussed.

The prevention of dysbiosis and pathogen proliferation through dietary strategies has been approached by several means, usually in combination. The following methods will be discussed: increasing feed intake pre-weaning by creep feeding (4.1); directly reducing pathogen proliferation, that is, with compounds or feeds containing antimicrobial activity (4.2); promoting the growth of beneficial microbial members that would then limit pathogen proliferation (4.3); and modulating the substrates available to the microbiota by reducing the level of fermentable dietary protein (4.4).

5.1 Creep feeding

The frequently observed anorexia/very low feed intake in the immediate days following weaning is a key factor impairing the gut microbiota - leading to dysbiosis and the associated increased risk of inducing PWD (Guevarra et al., 2019, Lalles et al., 2004). Therefore, strategies to keep a relatively high feed intake post-weaning have been proposed. A common practice to stimulate feed intake post-weaning is offering creep feed, that is, offering feed to the piglets while suckling. This would enable the gut ecosystem to adapt, including the microbiota, to cope with substrates, for example, of plant origin, other than those originating from the sow's milk. The animals are introduced to plant substrates while still receiving immunological protection and substrates from the dam's milk to which their digestive system is adapted to at that life stage.

The hypothesis is that the transition to receiving only feed after weaning would be less abrupt and would therefore reduce the detrimental impact on the microbiota composition and digestive system. One of the implications of this strategy is 'enzyme training,' that is, the digestive enzyme activity is stimulated to hydrolyse substrates of plant origin in a more progressive manner, so that the digestive system is relatively adapted to the feed at weaning. Similarly, the microbiota community would also adapt and therefore evolve more progressively to ferment plant substrates, without resulting in a dysbiotic situation post-weaning. This progressive adaptation of gut function also occurs in nature but over a much longer period.

The impact of consuming creep feed on post-weaning feed intake and body weight gain has been observed in several studies, in which piglets were classified as eaters or non-eaters pre-weaning. Eaters had a higher feed intake and body weight gain during the post-weaning weeks compared with non-eaters or those not offered creep feed (Bruininx et al., 2002, Bruininx et al., 2004, Kuller et al., 2007, Sulabo et al., 2010). Further, Kuller et al. (2007) reported higher net absorption in the small intestine of eaters than of that of non-eaters and suggested that creep feeding could be a useful tool in the prevention of PWD. This was studied by Carstensen et al. (2005), who found a relationship between the low consumption of creep feed and low colibacillosis compared with a high or no creep consumption, resulting in a lower feed intake over the first two days post-weaning and a subsequent higher intake. These results are indicative of a balance between different factors in order to reach the optimal output. Unfortunately, none of these studies have investigated the impact of creep feed intake on the gut microbiome.

Several dietary ingredients or additives such as highly digestible milk-derived products, sweeteners, plant extracts and oils, or feeding strategies like feeding with liquid feed/fermented liquid feed (FLF), are some of the practices applied that aim to stimulate feed intake pre-weaning, and as a consequence, post-weaning. It is beyond the scope of this chapter to describe how feed intake by piglets can be stimulated by such dietary ingredients and additives. However, it seems likely that the microbiota may be influenced by such dietary interventions.

5.2 Antimicrobial activity

When aiming to reduce the growth of *E. coli*, one of the main aetiologic agents of PWD, the antimicrobial activity of various strategies is achieved by reducing the pH and increasing the concentration of acids in the gut lumen. We have selected some examples of these, which have not been described in other chapters, for example, organic acids, FLF and some plant materials. The aim of these strategies (although they can also have other effects, here the focus is the antimicrobial impact), is a bactericidal, or at least bacteriostatic effect, against *E. coli*.

5.2.1 Organic acids

Several studies (Kirchgessner et al., 1992, Canibe et al., 2001, Knarreborg et al., 2002b, Canibe et al., 2005, Kluge et al., 2006, Suiryanrayna and Ramana, 2015) have reported the antibacterial effect of organic acids against Enterobacteriaceae and, when administered at high dose, also against other microbial groups, such as lactic acid bacteria, anaerobes *per se* and yeasts. Variations in the observations were found, as reviewed by Luise et al. (2020). For example, higher levels of lactic acid bacteria at lower levels of organic acid inclusion have been reported (Li et al., 2007, Luise et al., 2020). Some studies have also shown that organic acid addition resulted in reduced PWD (Tsiloyiannis et al., 2001, Bosi et al., 2007, Partanen et al., 2007, Papatsiros et al., 2011, Callegari et al., 2016) as well as reduced *Salmonella* shedding, and lower counts of *Salmonella* and *E. coli* have been observed in the stomach content of piglets experimentally challenged with these pathogens (Taube et al., 2009, Barba-Vidal et al., 2017). The addition of organic acids to the feed of young pigs is currently common practice in many countries.

The antimicrobial capacity of various organic acids differs (Knarreborg et al., 2002b) and is related to the reduction of pH, as well as the ability to dissociate (generating antimicrobial anions), which is determined by the pKa value of the respective acid and the pH of the surrounding milieu (Cherrington et al., 1991, Russell, 1992). An antimicrobial effect can be seen without a pH reduction, for example, when added as a salt (Canibe et al., 2001).

The impact of organic acid addition to the feed on the microbiota profile is much less studied, but some older studies, based on DGGE (denaturing gradient gel electrophoresis) and T-RFLP (terminal restriction fragment length polymorphism), have shown a reduced microbial diversity in the upper gastrointestinal tract as a result of feeding organic acids in blends or as a single acid, such as formic acid (Namkung et al., 2004, Canibe et al., 2005). Conversely, Roca et al. (2014), detected an increased microbial diversity in the colon of weaners using T-RFLP after feeding with 0.3% sodium butyrate; similarly, Torrallardona et al. (2007) observed a higher diversity in the ileum after adding 0.5% benzoic acid to the diet and using RFLP. Luise et al. (2017) also measured a higher alpha diversity (Chao 1) in the jejunum after feeding weaners diets with 0.64% formic acid but not with 0.14%; whereas Sun et al. (2020) reported lower diversity (Chao1) in the caecum of growers fed with 0.2% sodium butyrate. These nuances complement the general view of promoting high microbial diversity in order to increase gut health or reduce the risk of PWD. These inconsistent results also seem to indicate that factors like dose, treatment duration, chemical form, intestinal site, type of diet and age of the animal may also influence the effect of dietary organic acids fed to piglets on the gastrointestinal microbiota (Luise et al., 2020).

In the study of Luise et al. (2017), although a ß-diversity analysis (NMDS (non-metric multidimensional scaling) plot on the Bray–Curtis distance matrix) did not clearly distinguish clusters based on diet, some differences in the microbiota profiles were detected. Acid addition resulted in a decreased abundance of both lactic acid bacteria (*Lactobacillus*, *Leuconostoc* and *Gemella*) and butyric-acid-producing bacteria (*Fusobacteria*). Sun et al. (2020) measured a reduction in the relative abundance of *Firmicutes*, *Proteobacteria* and *Synergistetes* in the caecum content, and an increase in the abundance of *Bacteroidetes* and *Tenericutes* with dietary sodium butyrate.

When feeding organic acids, there is a higher concentration of the corresponding acid (compared with the non-acid-supplemented control diet) in the proximal gastrointestinal tract, that is, the stomach and small intestine; however, no differences are seen in the hindgut (Canibe et al., 2001, Canibe et al., 2005) as the acids are absorbed in the small intestine. Reaching a high concentration of the acid in the proximal gastrointestinal tract is desirable, as killing acid-sensitive pathogens (such as *E. coli*) proximally would reduce their numbers more distally and would lower the risk of disease along the gastrointestinal tract and of transmission via faeces. There are examples where the objective is to allow the added organic acid to reach the more distal segments of the gut, for example, butyric acid, which is an important energy source for the colonocytes, thereby contributing to a well-functioning colon epithelium (Guilloteau et al., 2010). Coating/microencapsulating technologies are used to protect the acids that are targeted at a high concentration more distally, where they can then be released (Mallo et al., 2012, Barba-Vidal et al., 2017, Yang et al., 2017, Abdelli et al., 2020). Another technology to deliver organic acids to the distal gut is based on butyrylation, that is, the introduction of butyryl groups into a substance/ingredient, for example, high-amylose maize starch, as studied in rats by (Clarke et al., 2008, Nielsen et al., 2019). By feeding these modified ingredients, a higher concentration of butyric acid in the colon and beneficial effects in relation to colorectal cancer in humans have been measured (Clarke et al., 2008, Nielsen et al., 2019).

As summarised by Canibe (2019), organic acids have long been shown to positively affect health and growth parameters in young pigs, mainly due to their antimicrobial effect and the impact on protein digestibility. More recent studies have aimed to identify other modes of action related to epithelial integrity and immune response around weaning, with the results showing some effects. There are few studies, some of which have used combinations of acids with other components, which makes interpretation more difficult. Moreover, although some effort has been made to separate the possible effects of organic acids through the known impact on the microbiota from the more direct effects of the acids, this aspect needs further elucidation.

Mixtures of organic acids and essential oils, botanicals, medium-chain fatty acids or other compounds are increasingly being tested with the purpose of obtaining synergistic effects to modulate the microbiota, the intestinal mucosal barrier and immune response (Grilli et al., 2015, Kuang et al., 2015, Ferrara et al., 2017, Li et al., 2018a, Pu et al., 2018, Xu et al., 2018, Jimenez et al., 2020). This approach is in line with the belief that due to the complexity of the processes occurring at weaning, various parameters need to be regulated around weaning in order to provide robustness and resilience to the gastrointestinal ecosystem.

5.2.2 Fermented liquid feed (FLF)

The fermentation of pig feed ingredients or whole feed is a strategy aimed at various outcomes, including reduction of the level of antinutritional compounds, increased digestibility of nutrients and/or added biosafety by promoting lactic acid fermentation with the corresponding antibacterial capacity. Fermented feed ingredients are typically dry products manufactured by a specific company and added to the feed, as with any other ingredient, for example, fermented soybean products, and fed as dry feed, which will not be further discussed here. Conversely, liquid feed and fermented liquid feed (FLF) are most often prepared on the farm by mixing the feed ingredients either dry or in the form of liquid co-products and water.

The fermentation process starts when feed and water are mixed. The profile develops over time, depending on the temperature (a higher temperature accelerates the process), and ingredient composition. When fermenting a standard pig diet, the proliferation of coliform bacteria occurs initially, and starts to decline when the parallel growth of lactic acid bacteria results in a high concentration of lactic acid and a lower level of acetic acid, with a corresponding reduction of the pH (see Fig. 2) (Brooks, 2008, Canibe and Jensen, 2012). The addition of organic acids or starter cultures can help by avoiding the initial growth of coliform bacteria (Canibe and Jensen, 2012). This illustrates the importance of maintaining a controlled process when preparing FLF in order to obtain a good quality product.

The high level of organic acids, that is, lactic acid and acetic acid, and the low pH confer an antibacterial effect on FLF (Van Winsen et al., 2001a, Boesen et al., 2004, Lindecrona et al., 2004, Canibe and Jensen, 2012). This results in a bactericidal effect on *Salmonella* and coliforms in the feed (Beal et al., 2002, Canibe and Jensen, 2003, Canibe and Jensen, 2012) and in the gastrointestinal tract (Canibe and Jensen, 2003, Van Winsen et al., 2001b). The reduction of other pathogens such as *Lawsonia intracellularis* and *Brachyspira hyodisenteriae* have also been reported when feeding pigs with FLF (Boesen

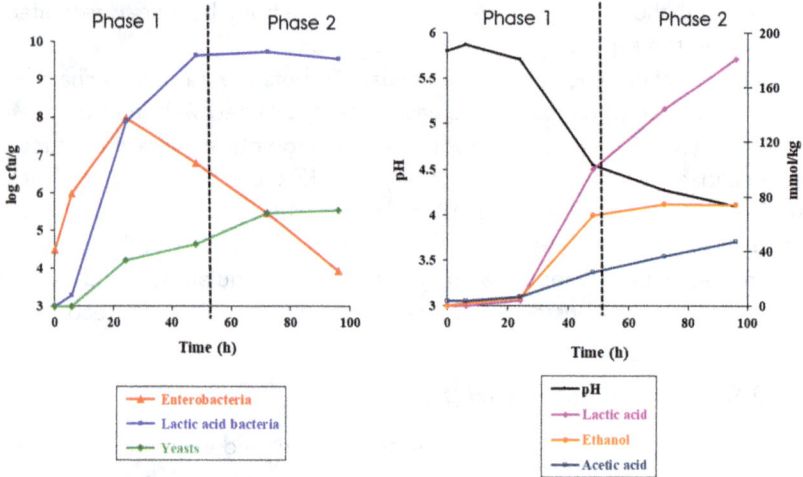

Figure 2 Illustration of the characteristics of liquid feed based on a standard grower diet and water during incubation at 20°C.

et al., 2004, Lindecrona et al., 2004). These antibacterial characteristics make FLF a relevant feeding strategy during the weaning period as a means of reducing the risk of pathogen proliferation.

Fermentation also offers the possibility of providing the animals with specific beneficial bacteria, or probiotics, in high numbers. Probiotic products, which are typically added to the dry feed at levels of around 10^9/kg feed, can be supplied at much higher levels if the probiotic strain is able to grow in the liquid feed during fermentation, and can reach values of up to ~1000 times higher, which is 10^9/g FLF (Van Winsen et al., 2001b, Canibe et al., unpublished). A specific probiotic added to dry feed at 10^6/g, that is then subjected to liquid fermentation, can reach levels of up to 10^9/g FLF (Canibe et al., unpublished). As one of the parameters influencing the impact of probiotics on the animals is the amount provided, FLF can increase the amount, and potentially, the effect of the probiotics on the host.

The addition of specific enzymes to the feed prior to liquid fermentation can also reduce the risk of imbalance in the gut during the post-weaning period by increasing the digestibility of plant substrates, which the piglet's system is not fully developed to cope with. Data can be obtained for growers (Jakobsen et al., 2015b, Jakobsen et al., 2015a) but studies with weaners are lacking in this respect.

Another important aspect when dealing with weaners is to avoid a period of anorexia immediately after weaning. As the piglets are obviously used to a liquid diet during suckling, it is hypothesised that feeding with liquid feed at weaning or even before weaning, so that they are already familiar with it at weaning, would help the piglets maintain a relatively high feed intake compared with dry feed (Russell et al., 1996, Missotten et al., 2015). Published studies that test this hypothesis are lacking. In general, the potential beneficial impact of feeding FLF of good nutritional and microbial quality around the weaning period is achieved by: protection from pathogen proliferation via its antibacterial effect; prevention of a period of anorexia due to a consistency similar to that of milk; provision of high numbers of probiotic strains; and contribution to improved nutrient digestibility by enzymatic activity pre-feeding.

The preparation of FLF can pose some challenges regarding both microbial and nutritional characteristics of the final product. Therefore, knowledge of the dynamics of the process is required in order to succeed in obtaining a good quality feed. These aspects have been reviewed elsewhere (Brooks, 2008, Canibe and Jensen, 2012, Missotten et al., 2015).

5.2.3 Plant materials

The supplementation of plant materials in weaner feed as dry whole plant material, as extracted compounds (e.g. essential oils), alone or in combination with other compounds (e.g. organic acids as described above) is also an extended strategy, aiming at achieving not only antimicrobial, but also immune-regulating and gut epithelial health-promoting effects (Omonijo et al., 2018, Van Noten et al., 2020, Yang et al., 2015).

Although the handling and dosing of extracted and varyingly purified compounds seems most convenient, it should be kept in mind that various interactions may affect their mode of action in the host plants. Therefore, the level and nature of the effects of whole plant material (e.g. leaves, bulbs or berries) or compound mixtures (cocktails) may differ considerably from that of the individual compounds.

The term 'essential oils' typically signifies composite plant extracts, often with a few dominating compounds (like the terpenes carvacrol and thymol in the essential oils of oregano and thyme), but may in total comprise dozens of different compounds; however, the term is also used for the individual compounds themselves, such as carvacrol or thymol (Bakkali et al., 2008, Omonijo et al., 2018).

The antimicrobial activity of essential oils is partly related to their hydrophobicity and hence their capacity to penetrate lipid membranes and mediate disruption of the cell membrane integrity and increased membrane permeability. However, essential oils may also affect microbial quorum sensing, resulting in impaired or inhibited toxin production or biofilm formation (Bouyahya et al., 2019, Li et al., 2019).

Cell membranes differ among bacteria with respect to lipid composition, which affects their sensitivity towards essential oils. In addition, the lipopolysaccharides in the outer membrane of Gram-negative bacteria make up a hydrophilic barrier against the penetration of hydrophobic compounds (Trombetta et al., 2005). The outer membrane of Gram-negative bacteria is a target for the action of essential oils, as observed by the release of lipopolysaccharides upon exposure to essential oils (Helander et al., 1998), and Gram characteristics *per se* do not determine bacterial sensitivity towards essential oils (Bouyahya et al., 2019, Trombetta et al., 2005). Moreover, and as suggested by Bouyahya et al. (2019), since essential oils contain a variety of chemical constituents, it seems probable that more than one mechanism is responsible for their antibacterial activity. However, *in vivo* studies with weaning piglets have indicated that the mode of action of essential oils may include promotion of commensal Gram-positives, for example, *Lactobacillus* sp., and/or suppression of potential pathogenic Gram-negatives, in particular *E. coli* (Castillo et al., 2006, Namkung et al., 2004, Wei et al., 2017).

As for organic acids, products based on microencapsulated essential oils have been developed with the aim of conferring stability on the bioactive compounds during feed processing and passage through the upper gastrointestinal tract, and to enable a slow release along the gastrointestinal tract (Choi et al., 2020, Abdelli et al., 2020). A recent attempt to encapsulate or at least protect thymol via gluco-conjugation was not successful, and the unprotected – not the protected – form of thymol was observed to reduce the piglet faecal score and diarrhoea incidence; in this case, neither of the two formulations affected the microbiota composition (Van Noten et al., 2020).

As described in Section 5.2.1, the synergistic effects of essential oils and organic acids have been studied (Abdelli et al., 2020 Grilli et al., 2015) and it has been suggested that the essential oils may render the cytoplasmic membrane more penetrable to organic acids, which can then gain easier access into the target bacteria (Omonijo et al., 2018).

In line with this, we have conducted *in vitro* studies testing the antibacterial effect of various plant materials, alone and in combination, against ETEC fimbriae types F4 and F18 in gastric and intestinal digesta, and demonstrated synergistic effects (Højberg et al., unpublished) (Fig. 3). We observed that a combination of ramsons and acid berries (lingonberries or redcurrants) had a more potent antibacterial effect than each plant material separately. The berries contribute by providing organic acids and the ramsons have a high content of allicin, known for its antimicrobial activity (Salehi et al., 2019). However, the exact mechanism behind the observed synergistic effect is not clear. We chose to study the non-extracted plant material of ramsons because allicin is an unstable compound, and not, for example, the essential oils of garlic (a close relative of ramsons) because these are mainly made up of allicin degradation products and may exhibit less antimicrobial effect (Stojanović-Radić et al., 2013). A pilot *in vivo* study with piglets showed a clear reduction in the number of Enterobacteriaceae, but

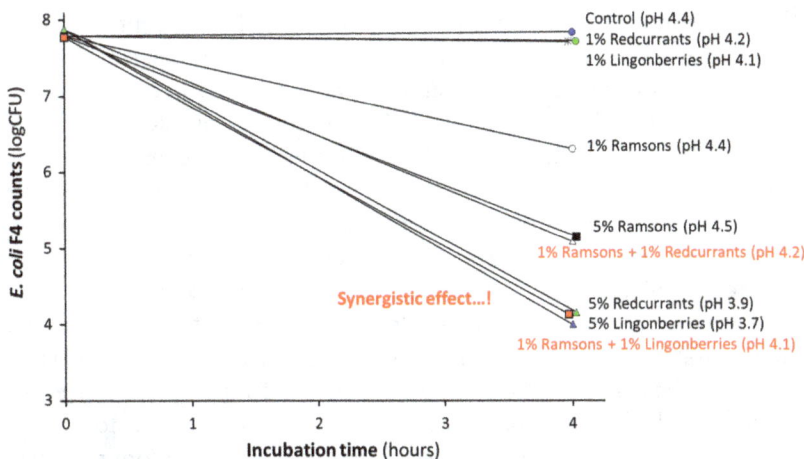

Figure 3 *In vitro* survival of a strain of *E. coli* fimbriae type F4, inoculated into pig stomach contents (initial pH 4.4, final pH values are indicated in parenthesis after each treatment), treated with ramsons and/or acid berries (redcurrants or lingonberries). The plant material was added as freeze-dried powder of bulbs and berries at levels of 1% or 5% (w/w). There was a clear dose-response for the individual plants. Further, combining 1% ramsons and 1% berries gave more than an additive effect, indicating synergistic interactions.

not of lactic acid bacteria in faeces and along the gastrointestinal tract, when piglets were fed with a standard weaner diet supplemented with 3% ramsons and 3% lingonberries, included as freeze-dried powder of bulbs and berries, compared with the same non-supplemented diet (Canibe et al., unpublished).

This type of result illustrates the potential of compositing additives by combining compounds with different modes of action into more effective products. However, it requires further knowledge on these modes of action to make qualified decisions on the compounds expected to provide synergistic or additive outcomes and thus are to be combined.

5.3 Probiotics, prebiotics and synbiotics

5.3.1 Probiotics

A large number of original publications and reviews are available in the literature on probiotics, dealing with their proposed mode of action and their impact on the parameters of host health and performance (Barba-Vidal et al., 2017, Dubreuil, 2017, Liao and Nyachoti, 2017, Sanders et al., 2018).

Probiotics are defined as live microorganisms that, when administered in an adequate amount, confer a health benefit on the host (Hill et al., 2014). In pigs, the probiotics most commonly used are lactic acid bacteria, bifidobacteria, bacilli and yeasts (Barba-Vidal et al., 2017, Dubreuil, 2017, Liao and Nyachoti,

2017). Their modes of action are considered to include the production of antimicrobial substances, competition with pathogens for adhesion sites and nutrients, enterotoxin inactivation, enhancement of mucosal barrier integrity and immune modulation (Dubreuil, 2017, Roselli et al., 2017). Common statements on the impact of probiotics on pigs include: 'a consistent number of studies showed the potential capacity in terms of immunomodulatory activities of these feed additives in pigs, but contrasting results can also be obtained from the literature. Reasons for this are not clear but could be related to differences with respect to the probiotic strain used, experimental settings, diets, initial microbiota colonization, administration route, time and frequency of administration of the probiotic strain and sampling for analysis' (Roselli et al., 2017); 'from the literature it can be seen that, depending on the products used and the animal husbandry practices applied, feeding probiotics to pigs can improve pig gut health, nutrient digestibilities and growth performance' (Liao and Nyachoti, 2017); 'there is an increasing amount of scientific publication supporting that probiotic effects in gut ecology and/or immune stimulation may provide support to keep animals healthy. Another apparent aspect of the reported results is that it is extremely difficult to discuss and extract conclusions with the data reported to date because the conditions in which the probiotics have been tested are highly variable' (Barba-Vidal et al., 2017); 'from the recent studies performed using cell models and in the best cases confirmed using piglets, we can conclude that probiotics have the potential to play a positive role on ETEC (F4)-induced diarrhoea. Nevertheless, some inconsistencies in the results obtained in various studies constitute a bias that makes these data sometimes irreconcilable. For example, the choice of the probiotic strain, the age of the animal, the time of administration and the dosage are all relevant for the favourable outcome of a treatment' (Dubreuil, 2017). The general conclusion on feeding probiotics to piglets is that there are positive effects on various parameters related to health and performance but the results are variable and dependent on many factors that have not been fully identified yet.

One of the factors related to the importance of applying strategies before weaning is the timing of exposure to the probiotic. The so-called 'windows of opportunity,' meaning the periods in the piglet's life during which the chances of manipulating the system are the highest, are an important factor. Applying the probiotic as early as possible in the animal's life would have the highest chance of changing the microbiota, since the neonate has an unstable gut microbiota which is expected to be more receptive to manipulation. Providing probiotics directly to the piglets in the immediate post-natal period could be a possibility but no literature is available. Another way of supplying the neonatal piglets with probiotics is by adding them to the sows' feed. The positive impact of this strategy on the piglets has been reported (Alexopoulos et al., 2004, Kritas et al., 2015, Mori et al., 2011, Taras et al., 2006, Veljovic et al., 2017).

Application of a single strain versus a mixture or cocktail is another aspect of discussion. Whereas a combination of probiotic strains can be expected to supplement each other, for example, regarding mode of action, thereby potentiating the beneficial impact on the animal and has frequently been seen to be more effective than single strains (Chapman et al., 2011), issues of competition among the strains resulting in a reduced final impact have been considered (Chapman et al., 2011). This reflects the thorough knowledge on effects, mode of action, interactions etc., required before a probiotic product with the most beneficial impact can be designed.

Although there seems to be consensus on the fact that different probiotics have differing modes of action and therefore, not all probiotics can be expected to exert the same impact on the host, there are mechanisms that are shared among probiotic taxonomic groups at levels higher than strain (Sanders et al., 2018). Therefore, some general probiotic benefits can be expected in a non-strain-specific manner (Sanders et al., 2018). A clear example of a common mode of action of probiotics within the lactic acid bacteria group could be the bactericidal effect against pathogens via the production of lactic acid. On the other hand, production of a bacteriocin with a specific target can be strain specific (Sanders et al., 2018). Hence, in some cases, the traits may be shared broadly among different strains, whereas in other cases, it may be much narrower.

The microbiota composition of the host as a factor affecting the responsiveness to a probiotic has been considered in humans and has led to the term 'personalized probiotics' (Celiberto et al., 2018, Kort, 2014, Lee, 2018), that is, the benefits acquired from a probiotic are personal, depending on the health status, dietary habit and the prevailing gut microbiota of the recipient (Lee, 2018). Since probiotics interact with the host microbiota, the profile of this microbiota can instinctively be expected to affect the outcome of such an interaction. The topic of responders and non-responders to probiotics has drawn considerable attention, but little is known regarding why some individuals respond to a probiotic intervention and others do not (Reid et al., 2010). There is high individual variation in the gastrointestinal microbiota composition in pigs, as in humans, and therefore a different response to the same probiotic can be expected among individuals depending on their health status, the environmental conditions, genetics, feed composition and age etc. (Starke et al., 2013) provided sows with a probiotic strain of *Enterococcus faecium* and classified the recipient sows as responders or non-responders according to whether the gut microbiota of the individual was affected by the probiotic. Gaining knowledge on the influence of gut microbial profiles of the host on the responsiveness to specific probiotics would allow progress in matching probiotics and recipients, eventually leading to a higher effectiveness of this strategy.

5.3.2 Prebiotics

Prebiotics are defined as 'selectively fermented ingredients that allow specific changes, both in the composition and/or activity of the gastrointestinal microflora that confer benefits upon host well-being and health' (Gibson et al., 2010). According to the authors, although many dietary prebiotic candidates exist, the strongest data to date are reported for fructans, galactans and lactulose (Gibson et al., 2010). The focus so far has been on the impact on bifidobacteria and lactobacillus. Bindels et al. (2015) proposed revising this definition to reduce the focus on 'selective' or 'specific' and to add ecological and functional features of the microbiota more likely to be relevant for host physiology: 'a nondigestible compound that, through its metabolization by microorganisms in the gut, modulates composition and/ or activity of the gut microbiota, thus conferring a beneficial physiological effect on the host.' Given the network of interactions among microorganisms within the gastrointestinal tract, it seems most likely that more than one single bacteria group/genus/strain is stimulated by a prebiotic substrate. Next-generation techniques that have enabled moving from investigating a limited group of microorganisms to a more complete microbial profile, have contributed to motivate a new definition of the term prebiotic. An immediate effect of this proposed definition would be the inclusion of all non-digestible carbohydrates that improve health through a modulation of the gut microbiota as prebiotics (Bindels et al., 2015). Candidate prebiotics following this definition were also proposed: resistant starch, pectin, arabinoxylan, whole grains, various dietary fibres and non-carbohydrates that exert their action via modulation of the gut microbiota (Bindels et al., 2015, Hutkins et al., 2016). In the review of Tran et al. (2018), for example, a wide range of substrates are termed prebioticsAs discussed by Hutkins et al. (2016), who also highlighted the importance of having a common definition, there is less consensus on the definition of prebiotics than probiotics. A consensus among scientists on the most appropriate definition of a prebiotic is necessary to enable continued use of the term (Hutkins et al., 2016). In 2017, The International Scientific Association for Probiotics and Prebiotics (ISAPP) proposed a new definition: 'a substrate that is selectively utilized by host microorganisms conferring a health benefit' (Gibson et al., 2017). This definition expands the concept of prebiotics to possibly include non-carbohydrate substances, applications to body sites other than the gastrointestinal tract, and diverse categories other than food.

The impact of prebiotics on the piglet gastrointestinal ecosystem has been extensively studied and varying results have been reported. Although feeding weaners 4% fucosylated oligosaccharides (FOS) increased the molar proportion of butyric acid in the caecum and proximal colon, neither FOS nor 4% transgalactooligosaccharides (TOS) had a strong prebiotic effect on bifidobacteria along the gastrointestinal tract (Mikkelsen and Jensen, 2004). Similarly, no effect of the dietary addition of FOS or TOS (4%) was detected on the faecal number of bifidobacteria (Mikkelsen et al., 2003). None of the studies detected an effect of the prebiotics on total culturable anaerobic bacteria, lactic acid bacteria, lactobacilli, coliform and lactose-negative enterobacteria, whereas a stimulating effect on yeasts was observed in both studies (Mikkelsen et al., 2003, Mikkelsen and Jensen, 2004). Further, Mikkelsen et al. (2003) showed that several different species of bacteria present in the faeces of piglets are able to ferment FOS. Supplying FOS (10 g daily) at an early age, that is, from day 2 to day 14 of life, to suckling piglets, Schokker et al. (2018) measured a bifidogenic effect in the colon as well as changes in mucosal gene expression in the jejunum related to intestinal barrier function and immunity. Increasing amounts of dietary GOS resulted in reduced numbers of *E. coli* and higher numbers of *Lactobacillus* and bifidobacteria in the faeces of weaners (Xing et al., 2020). The results of a meta-analysis showed no effect of dietary inulin on ileal bifidobacteria and lactobacilli abundance, a negative relationship between dietary inulin and bifidobacteria and enterobacteria in the colon, and a negative relationship with lactobacilli and enterobacteria in faeces, with a tendency to a positive relation with bifidobacteria (Metzler-Zebeli et al., 2017).

A review of the effects of potential prebiotics in controlling *Salmonella* and *E. coli* discussed the various factors by which prebiotics can prevent invasion by these pathogens but pointed to the fact that many results were obtained using *in vitro* models (Tran et al., 2018). Some of the *in vivo* studies investigating the ability of prebiotics to protect piglets from infection by *Salmonella* and *E. coli* are those by Guerra-Ordaz et al. (2014), Rodríguez-Sorrento et al. (2020) and Letellier et al. (2000). Pig models in which weaners were challenged with *Salmonella* exhibited lower intestinal colonisation by this pathogen after feeding with an oligofructose-enriched inulin (dietary inclusion of 5%) (Rodríguez-Sorrento et al., 2020), and a reduction in shedding after supplementation with FOS (1%) in drinking water, although not when given in the feed (Letellier et al., 2000). Feeding ETEC F4 experimentally challenged weaners with a diet supplemented with 1% lactulose increased the lactobacilli counts, the butyric molar ratio in the colon and the villous height in the ileum, and reduced the acute-phase protein Pig-MAP in serum, but did not affect the colonic counts of enterobacteria, ETEC F4 or diarrhoea incidence in the piglets (Guerra-Ordaz et al., 2014). The addition of 8% inulin improved the faecal consistency and reduced the incidence of PWD in weaners experimentally infected with ETEC F4 (Halas et al., 2009).

5.3.3 Synbiotics

The definition of synbiotic has been updated by the International Scientific Association for Probiotics and Prebiotics, to 'a mixture comprising live microorganisms and substrate(s) selectively utilised by host microorganisms that confers a health benefit on the host' (Swanson et al., 2020). Two subsets of synbiotics were defined: complementary and synergistic. A 'synergistic synbiotic' is a synbiotic in which the substrate is designed to be selectively utilised by the co-administered microorganism(s) (Swanson et al., 2020). A 'complementary synbiotic' is a synbiotic composed of a probiotic combined with a prebiotic, which is designed to target autochthonous microorganisms (Swanson et al., 2020)

5.4 Dietary fibre

In general, there is conflicting evidence as to whether dietary fibre (DF) exerts a beneficial or detrimental influence on PWD. Insoluble fibre can reduce colonisation by *E. coli* and the severity of PWD (Gerritsen et al., 2012, Molist et al., 2014), whereas viscous soluble fibre can have the opposite effect (Hopwood et al., 2004, McDonald et al., 1999). Importantly, insoluble fibre *per se* leads to detrimental effects, with the high viscosity of some soluble fibre sources increasing the risk of PWD in piglets (McDonald et al., 1999, Hopwood et al., 2004, Wellock et al., 2008b).

Soluble fibres, especially viscous ones, can reduce the transit time of digesta through the small intestine, increasing the time taken for the digesta to reach the caecum; whereas insoluble fibre has the opposite effect on the transit time (Molist et al., 2014, Knudsen et al., 2016). A longer retention time in the small intestine would allow proliferation of fast-growing *E. coli*, thereby increasing the risk of PWD development. In other words, a prolonged transit time could allow the establishment of pathogens that would otherwise have been flushed through to the large intestine (Hopwood et al., 2004).

5.5 Reduced dietary crude protein (CP) level

The available dietary substrates are key factors in determining the composition of the microbiota along the gastrointestinal tract. The main components of a pig's diet, carbohydrates and protein, are metabolised by the gut microbiota to differing extents, depending on: (i) the digestibility of these compounds, leaving different amounts of substrates available for the microbiota and (ii) their fermentability, that is, the ability of the microbial community to metabolise them, which is affected by the physical and/or chemical characteristics of the substrates.

The microbial saccharolytic and proteolytic activity in the gut result in different metabolites with differing impacts on the host (Macfarlane and Macfarlane, 2012, Pieper et al., 2016, Gilbert et al., 2018) (Fig. 4a and b). While short-chain fatty acids, a result of both carbohydrate and protein fermentation, are considered to be health-promoting compounds, metabolites specifically produced as a result of protein fermentation, namely, ammonia, biogenic amines, hydrogen sulphide, indoles, p-cresol, branched-chain fatty acids and phenols are considered harmful to the host when present at high concentrations (Macfarlane and Macfarlane, 2012, Pieper et al., 2016). Although the exact mechanism behind the effect is not fully established, reducing the level of dietary protein in weaner diets leads to a reduced risk of PWD (Heo et al., 2008, Heo et al., 2009, Kim et al., 2011b, Pieper et al., 2016), which is probably due to the reduction in fermentable protein, hence reducing harmful metabolites that are in direct contact with the colonic mucosa and directly interact with the mucosal cells.

PWD is often related to proliferation of ETEC in the small intestine. However, it is known that non-infectious diarrhoea occurs during the initial two weeks post-weaning (Callesen et al., 2007) and the following weeks (Weber et al., 2015). The deleterious effects of high concentrations of metabolites from microbial proteolytic activity could explain the non-infectious type of PWD. Lower concentrations of metabolites from microbial proteolytic activity have frequently been reported after feeding low-protein diets compared with high-protein diets, mainly ammonia, biogenic amines and branched-chain fatty acids, in the ileum or caecum/proximal colon, with values of crude protein (CP) varying between 15% and 22% (Bikker et al., 2006), 17% and 23% (Nyachoti et al., 2006), 18% and 24% (Heo et al., 2008), 20% and 24% (Htoo et al., 2007), 17.5% and 22.5% (Opapeju et al., 2009), ~15% and 29% (Pieper et al., 2012a) and ~2% and 18% (Pieper et al., 2014).

On the other hand, high levels of CP exacerbate the incidence of diarrhoea in ETEC-challenged weaners (Wellock et al., 2008a). This could be due to a higher luminal pH in the small intestine as a result of the buffering capacity of protein, which supports *E. coli* growth, compared with more acidic conditions (Nyachoti et al., 2006) or more substrates available for this proteolytic bacterium (Heo et al., 2013).

Proteobacteria possess the broadest gene coverage of amino acid reactions, although only 9% are unique to this phylum, which is due to the high functional redundancy in the microbiome (Diether and Willing, 2019). The major amino-acid-fermenting bacteria in the gastrointestinal tract include proteolytic members of *Fusobacteria*, *Proteobacteria* (including *E. coli*), *Firmicutes* (including *C. perfringens*, *C. difficile*, *Selenomonas*, *Peptostreptococcus* and

(a)

(b)

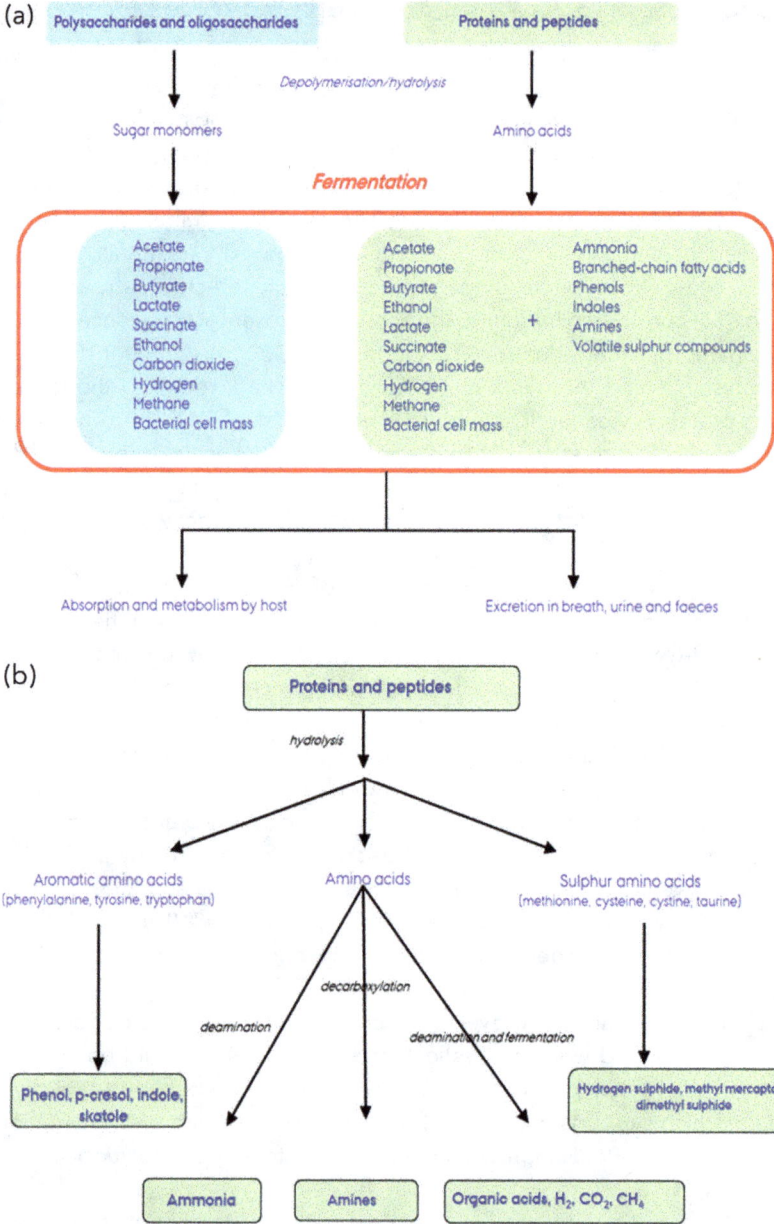

Figure 4 The main metabolites produced from microbial saccharolytic and proteolytic activity in the gut.

Veillonella) and *Bacteroidetes* (including *Bacteroides*) (Dai et al., 2011). An overload of undigested protein can trigger an overgrowth of some of these bacteria, and since some are pathobionts, it may also partly explain why high-protein diets in pigs have been associated with diarrhoea.

The impact of dietary protein on gut microbiota composition of piglets has, to date, mainly been explored by focusing on specific bacterial groups as opposed to the microbiota profile using NGS techniques. Here are some examples of these studies when reducing the dietary CP level compared with a control diet (reviewed by (Canibe, 2019): no effects on counts of lactobacilli or coliforms/Enterobacteriaceae were observed in the small intestine and/or colon (Bikker et al., 2006, Nyachoti et al., 2006, Hermes et al., 2009, Opapeju et al., 2009); no effect on total bacteria, Enterobacteriaceae, Bacteroides or the *C. coccoides* group in the proximal colon (Pieper et al., 2012a); no effect on T-RFLP profiles in the colon (Opapeju et al., 2009); increased numbers of the *C. leptum* group (Pieper et al., 2012a); increased lactobacilli-to-coliforms ratio (Wellock et al., 2006); no effect on coliforms in the ileum or proximal colon and decreased lactobacilli numbers in the colon (Wellock et al., 2008a, see also the review by Rist et al., 2013).

A few studies have addressed the impact of reducing the dietary CP level on the microbiota profile using 16S rRNA gene amplicon sequencing. Rattigan et al. (2020) reduced the level of dietary protein of a weaner diet from 21% to 18% and observed an increased in the relative abundance of Enterobacteriaceae and Helicobacteriaceae and a concomitant increased bacterial diversity in the colon digesta. Two other studies involving finishers (Fan et al., 2017, Zhou et al., 2016) did not show a clear relationship between dietary CP level and a 'healthier' microbiota composition in the gut. This illustrates the need for more studies on the relationship between PWD, microbiota composition, including pathogen abundance, and microbial metabolite concentration in the gut lumen, in order to elucidate the mechanisms responsible for the beneficial impact of reduced dietary CP level on the risk of PWD. Further, and equally as important, the correct diagnosis of diarrhoeic piglets, distinguishing infectious pathology from non-infectious, would allow a more targeted treatment resulting in higher effectiveness of this treatment and a lower use of antibiotics, which is a global agenda.

High levels of CP can negatively affect the epithelial barrier and histology but inconsistent results have been reported (Nyachoti et al., 2006, Hermes et al., 2009, Opapeju et al., 2009). However, dietary protein levels that are too low can also impair epithelial gut integrity (Fan et al., 2017), which is probably due to the limited bioavailability of specific amino acids in low-protein diets and is of major importance for host epithelial integrity.

As amino acids are essential for the animal, it is evident that a balance between the positive impact of reducing the level of dietary protein and the potential negative consequences, such as suboptimal growth, impaired gut integrity and immune system etc., have to be carefully considered.

Summary of dietary strategies: combinations of various additives or strategies in order to manipulate different processes or potentiate a specific process are gaining importance, which is in line with the complexity of the changes/alterations occurring in the animal's gut and beyond, such as the other organs including the brain, as a consequence of abrupt weaning.

6 Host factors influencing gut function

We have so far described the factors of importance for the prevention of pathogen colonisation and how dietary factors can influence the microbiota composition and function. To elicit infectious diarrhoea in piglets, ETEC must adhere to specific porcine enterocyte receptors for the respective fimbriae (Heo et al., 2013, Fairbrother et al., 2005). Fimbriae designated F4 and F18 are commonly found on ETEC in pigs with PWD. This adhesion facilitates colonisation of the distal jejunum and proximal ileum mucosa by the ETEC and enables them to transmit enterotoxins. Hence, any attempt to prevent adhesion of ETEC to the epithelium should prevent PWD. The ETEC are characterised by having two types of virulence factors: (1) adhesins, which allow their binding to and the colonisation of the intestinal epithelium and (2) enterotoxins, causing fluid secretion. The fimbriae are thin and flexible structures with a diameter of about 2-4 nm, whereas pili are more rigid structures with a diameter of approximately 7-8 nm and have an axial hole (De Graaf and Mooi, 1987), and both contain an adhesive lectin subunit. The adhesions are expressed in the ETEC fimbriae and differ between ETEC F4 and ETEC F18. The biosynthesis of fimbria is influenced by several factors, and pathogenic bacteria only fully express virulence factors such as fimbria when the conditions are appropriate for adherence and subsequent colonisation (Verdonck et al., 2004). Frydendahl (2002) identified genes for F4 and F18 in 92.7% of all ETEC involved in PWD. Once the ETEC have adhered to and colonised the small intestine, they can produce enterotoxins, leading to diarrhoea. Two classes of enterotoxins are produced by ETEC F4 and F18: heat-labile (LT) enterotoxins and heat-stable (STa, STb and enteroaggregative heat-stable toxin 1 [EAST1]) enterotoxins, causing electrolyte and net fluid losses. In addition, some *E. coli* strains produce a Shiga-like toxin type II variant, which causes oedema diseases. It is beyond the scope of this chapter to describe the many putative receptors that have been identified for ETEC F4 and F18 adhesion and their serological variances and subunits, but a brief overview is provided by Luise et al. (2019), and a detailed description of ETEC pathogenesis has been provided by Nagy and Fekete (1999). The next step is to describe the host-related factors which can improve gut function and prevent PWD.

6.1 Host genetics

Host genetics is a major factor for the prevention of *E. coli* adhesion to the gut epithelium. Generally speaking, only pigs expressing/possessing the fimbrial receptors F4 or F18 on the intestinal enterocytes are susceptible to ETEC diarrhoea (Fairbrother et al., 2005, Nagy and Fekete, 1999); however, there is individual animal variability between pigs that can be partially explained by genetic mutations associated with the expression of specific receptors on the intestinal epithelium. A DNA marker-based test has been employed to identify whether pigs lack or possess the F4 or F18R receptor (Fairbrother et al., 2005). The Mucin 4 (*MUC4*) gene on porcine chromosome 13 has been proposed as a candidate gene for the production of the specific ETEC F4ab/ac receptor (Jørgensen et al., 2003, Jensen et al., 2006) while the α-(1,2)-fucosyltransferase (*FUT1*) gene localised on porcine chromosome six has been proposed as a key gene for controlling the expression of *E. coli* F18R (Bao et al., 2012). With the DNA-based marker test allowing genotyping for F4ab/ac ETEC resistance/susceptibility, three genotypes were observed and classified as resistant (RR), susceptible heterozygote (SR) and susceptible homozygote (SS). It should be noted that in another review (Hodgson and Barton, 2009), it was concluded that there is no definitive simple test for F4 resistance. In addition to the *MUC4* gene, the transmembrane mucin MUC13 appears to be significantly associated with susceptibility and resistance to F4ab/ac, as the *MUC13* gene is closely linked to the genes encoding F4ab and F4ac receptors (Goetstouwers et al., 2014). Likewise, regarding pig resistance to ETEC F18 infection, pig intestines also express *FUT2*, and polymorphism in *FUT1* genes influence FUT1 and FUT2 enzymes and may influence gut intestinal homeostasis and the commensal microbiota (Poulsen et al., 2018a).

Genetics alter susceptibility to ETEC adhesion, and breeding programs with F4 or F18 receptor-negative pigs would potentially result in prevention of F4ab/ac-induced (or likewise F18) ETEC diarrhoea in a pig population. However, although the Danish pig population is considered genetically resistant to F4 ab/ac due to the applied breeding programme, the incidence of diarrhoea has not been effectively reduced, according to the authors' knowledge. Targeting a multifactorial production disease such as PWD by selection based on single genes is probably not efficient, meaning that in addition to the receptors on the surface of the pig intestinal epithelium, scientists have studied other host factors influencing gut function in terms of disease resistance.

6.2 Physiological/developmental stage of the host

There are several physiological factors of importance for the development of gut function, and we will address how the physiological status and maturity of the pig influence its capability to cope with the challenges around weaning. Furthermore, we will discuss some manageable factors (vaccination, age at weaning, feed intake post-weaning) that can aid in better development of gut function (Mulder et al., 2009).

6.2.1 Barrier function

The gastrointestinal barrier comprises a multi-layered system of host defence mechanisms, provided by the intestinal epithelial cells and components of the immune and enteric nervous systems. The gut must effectively transport luminal nutrients, water and electrolytes, while providing a barrier to pathogenic and antigenic components in the lumen. An appropriate and effective immune response by the host is necessary to distinguish relatively harmless antigens (coming from food) and the commensal bacteria from potentially harmful antigens and external pathogenic species (pathogenic bacteria, toxins, viruses, parasites and fungi). The innate and acquired (also called adaptive) immune systems are functionally well coordinated to protect the host. Under normal conditions, the immune system acts precisely and promptly in a temporary action against invading pathogens. However, when the immune system response is prolonged and unresolved, inflammation persists, resulting in a situation known as chronic low-grade inflammation. The challenge is that pigs are commonly weaned at an age when their immune system is not mature. The presence of commensal bacteria has a direct influence on immune maturation and microbial exposure influences the expression of a large number of immune-related genes (Mulder et al., 2009). In commercial pig production, weaning is an abrupt process occurring between 14 and 30 days of age, while in nature, the weaning process occurs gradually and approaches completion at around 10 to 12 weeks or even up to 18 weeks of age. The timing of commercial weaning coincides with a period of declining passive immunity from the sow's milk, while the active immune system is still immature (Moeser et al., 2017). Thus, any inappropriate or excessive immune stimulation during this critical period may potentially disrupt the development and long-term function of the immune system. Studies have shown that functional gastrointestinal disturbances in the epithelial barrier, immune and nervous system functions in early-weaned pigs persist into adulthood (Medland et al., 2016, Pohl et al., 2017).

6.2.2 Vaccination

One of the obvious strategies to prevent E. coli infection and hence PWD would be vaccination. While neonatal infections in general can be effectively prevented by passive colostral and lactogenic immunity obtained by vaccination of the sow (Melkebeek et al., 2013), the challenge regarding ETEC vaccination of pigs has been to administer an effective vaccine that provides the required protection post-weaning, when lactogenic immunity disappears. At least one commercial vaccine (Coliprotec) for oral use against PWD is available; however, the efficacy of this vaccine is unknown. Scientists (Verdonck et al., 2007, Tiels et al., 2008) have looked into the development of mucosal vaccines, by which the local F4- and/or F18-specific sIgA response could be activated; however, such vaccines are not currently available commercially.

6.2.3 Weaning age

One of the major factors pinpointed in preventing PWD is the weaning age. Weaning at an early age is associated with the presence of weaning diarrhoea (Madec et al., 1998). In this cohort study, pigs above or below a weaning age of 26.5 days were compared and the occurrence of diarrhoea was 12 times higher in herds with a low weaning age. In addition, a weaning weight below 7.2 kg enhanced the occurrence of diarrhoea by 6.5 times compared with pigs weaned at a weight of 8-9 kg but was almost unchanged compared with pigs weaned at a weight of 7.2-8.1 kg (Madec et al., 1998). McLamb et al. (2013) showed that pigs aged 15-16 days at weaning had increased incidence of clinical disease (diarrhoea and growth performance reduction) upon challenge with F18 ETEC, increased intestinal permeability and a compromised or suppressed immune response when compared with pigs weaned at 22 days of age. Other studies have also shown that early-weaned piglets have compromised immune responses at a later age (Davis et al., 2006), and a study by Moeser et al. (2017) indicated that early-weaned pigs have marked and persistent mast cell hyperplasia in the small and large intestine when measured at two and nine weeks post-weaning. Mast cells are haematopoietically derived innate immune cells that play important roles in host defence and disease pathogenesis. The cells are located at host-microbiota interfaces such as the gut mucosa, and the cells are activated within 24 h of weaning in pigs, and have been suggested to play a central role in driving intestinal epithelial permeability disturbances in the early-weaned pig (Moeser et al., 2017). The stress that is associated with weaning manipulates the immune system, resulting in mast cell activation, which may result in loss of barrier function of the tight junctions in the small intestine (Wijtten et al., 2011). An increase in permeability can favour the passage of toxins and pathogens through the epithelium (Brown and Price, 2008). It remains to be elucidated how weaning weight is related to PWD incidence, as most of the studies included in the previous sections have been performed with indication of weaning age but no indication of weaning weight. Due to genetic development, the large litter sizes may reduce the average piglet weight at weaning; however, the influence on gut function is unclear.

6.2.4 Feed intake and undernutrition

Another factor of major importance for gut function is the level of feed intake after weaning, as also described above for the gut microbiota. The small intestinal barrier and absorptive functions deteriorate within a short time after weaning (Wijtten et al., 2011). The loss of barrier function due to low feed intake is due to a shortage of luminal nutrient supply. Burrin and Stoll (2003) divided the quick and marked alterations in the digestive and absorptive, barrier and immune functions into two phases, that is, an acute phase, observed within the first 5-7 days after weaning and a subsequent adaptive phase. These authors distinguished between the phases primarily on the changes in feed intake and the subsequent impacts enteral (luminal) nutrition has on the gut. If the gastrointestinal tract becomes deficient in macronutrients, micronutrients and energy, then health development and any subsequent recovery in the adaptive phase will be impaired (Pluske, 2013). It should be noted that immune defences are energetically expensive. Therefore, the rate at which pigs transform energy and nutrients can be expected to be elevated as a result of immune defence activation. While the costs of developing the immune system are considered generally low due to the lack of diversification process and low rates of cell turnover when an immune response is not being mounted, activation of the innate response is generally considered to be costly and even more costly than activation of the adaptive response (Rauw, 2012). One of the earliest responses to infection is cytokine-mediated anorexia, where interleukins 1, 6 and 8, TNF-α (tumour-necrosis factor) and INF-γ (interferon) are released by the host defence mechanisms – resulting in reduced nutrient intake through effects on the central nervous system (Donabedian, 2006). The immune system does not have to be challenged to a great degree to alter nutrient dynamics in the host because even rather mild immune reactions, like those associated with vaccination, can suppress feed intake and development (Lochmiller and Deerenberg, 2000). Protein and several other nutrients including vitamins, minerals and fatty acids are known to influence immune functions. Therefore, in theory, scarcity of any of these nutrients may cause reduced resistance to infection. The amino acid pattern required to support the immune response is different from that released by skeletal muscle proteolysis, resulting in an excess of non-limiting amino acids, whereas others become limiting for the immune system. Likewise, for micronutrients, their resource priority may increase when the pig enteric immune system is challenged (Lauridsen et al., submitted). Coop and Kyriazakis (1999) theorised that growing animals encountering parasites for the first time can be expected to prioritise resources to acquisition of immunity over growth, whereas once immunity has been acquired, growth would be prioritised over expression of immunity to parasites. It should be noted that genetic selection towards a superior growth rate may result in decreased resistance to disease or a reduced immunological response.

Thus, undigested protein could be present due to an activated immune system in pigs post-weaning, and when this escapes to the lower part of the small intestine, it may be even more harmful for the host as the intestinal barrier function is impaired due to the lack of luminal nutrients or other factors related to the weaning process. One solution would be to offer a low-protein diet during the initial days post-weaning supplemented with essential amino acids, including those amino acids of importance for immune responses and epithelial barrier function. However, clarification is needed regarding what amino acids are catabolised in the maintenance of a state of readiness and are involved in combatting infection and repair of damaged tissue caused by common weaning stress.

6.3 Importance of inflammatory responses and oxidative stress for post-weaning diarrhoea development

So far, we have described how the development of gut function can be influenced by genetic and environmental factors. We will now discuss how the development of clinical disease (i.e. diarrhoea) is highly dependent on the host's inflammatory response level, which has been investigated for human enteric inflammatory diseases including IBD (Caruso et al., 2020). Figure 5 describes the various steps during early and late dysbiosis leading to diarrhoea.

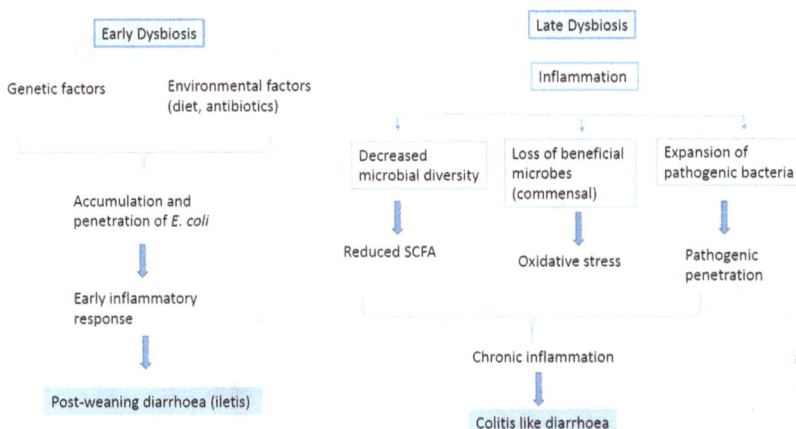

Figure 5 Summary of how inflammatory responses can lead to diarrhoea.

6.3.1 Early inflammatory response

- Local inflammation in the small intestine occurs during the weaning phase (weaning-associated intestinal inflammation).
- Genetic and environmental factors affect the accumulation and penetration of pathogenic bacteria.
- The increased production of proinflammatory cytokines reduces the feed intake and increases body temperature, and induces a shift in the metabolism to a more catabolic state which results in reduced weight gain and feed utilisation.
- This leads to adverse changes in gut function in terms of intestinal morphology, including reduced villus height, increased villus width, increased crypt depth and reduced absorptive capacity and brush border enzyme activity.
- Pathogenic bacteria, such as *E. coli*, encounter favourable growing conditions during this stage, known as colibacillosis (ETEC diarrhoea).

6.3.2 Chronic inflammation

- Dysbiosis leads to: decreased microbial diversity, loss of beneficial microbial communities and expansion of pathogenic microbial communities.
- The mucosal firewall is broken, that is, the luminal antigens pass into the lamina propria where an inflammatory response can be escalated.
- The production of SCFA is reduced, impairing the gut barrier.
- Intestinal mast cells are activated, which are involved in disruption of the intestinal barrier.
- Barrier integrity is disturbed, which would favour the transepithelial passage of various pathogenic bacteria etc. and promote bacterial infections and diarrhoea.
- Oxidative stress reactions develop during the highly activated mucosal firewall.
- Diarrhoea and other clinical symptoms related to the severity of the inflammation are developed.

The stress occurring around weaning may promote intestinal inflammation that can be exacerbated by enteric infections. During a gut inflammatory host response, reactive species such as nitric oxide develop, and when transformed into nitrate in the gut lumen, favour the growth of specific bacterial strains; in other words, the growth of *E. coli* is advanced, rather than that of *Clostridia* or *Bacteroidia*. There is, in general, a strong interaction between oxidative stress and inflammation. Immune cells, such as granulocytes and neutrophils exposed to bacterial cell wall peptides or lipopolysaccharides are part of the host inflammatory response. Thus, the generation of oxidants exerts an important role in killing pathogenic bacteria such as *E. coli*, but these oxidative stress reactions may also be harmful as the oxidants can damage components in host cells, and if these reactions are uncontrolled, the epithelial barrier is disrupted (Lauridsen, 2019). Hence, one mechanism to diminish inflammatory mediator production may be to prevent oxidative stress (Calder et al., 2009). This can be accomplished via enhancing antioxidant defence mechanisms (Lauridsen, 2019), which can be achieved by providing dietary antioxidative micronutrients (Lauridsen et al., submitted). Micronutrients are well known for their role in modulating the immune function, and several vitamins have both a direct and indirect influence on immune cells, with the majority of clinical research focusing on the relevance of human infectious diseases. We will not go into detail on micronutrient use as a strategic tool to enhance gut function; however, they are capable of modulating not only weaning-associated intestinal inflammation, but also enteric microbiota composition and function. One of the most researched micronutrients to prevent PWD is zinc, which deserves special focus.

6.4 Use of medicinal zinc to prevent post-weaning diarrhoea

As antibiotics can promote intestinal inflammation and dysbiosis, as well as loss of microbiota diversity, prophylactic use of antibiotics may be harmful for gut function. Thus, significant public health concern about the spread of multi-resistant bacteria and the role of antibiotics in impairing gut function has encouraged several scientists to search for alternatives to antibiotics. Zinc, or more specifically zinc oxide (ZnO), is added to nursery pig diets at 2000 mg/kg to 4000 mg/kg to alleviate diarrhoea and check the growth of the post-weaning piglets in many pig-producing countries. However, an increase of multi-resistant *E. coli* in association with ZnO feeding in piglets has been reported (Ciesinski et al., 2018). Since ZnO has been proposed as one of the most effective feed additives to replace antibiotics, one may ask how medical ZnO can improve gut function to prevent PWD.

Since the late 1980s, the practice of adding medicinal levels of ZnO to pig feed in the immediate period post-weaning has become widespread in commercial pig production in several European countries. Thus, the use of ZnO in Danish pig production increased markedly after the withdrawal of antibiotic growth promoters in the 1990s and peaked in 2015. However, the EU decision of 2017 to ban the use of medicinal ZnO from 2022 is reflected in a decreased use of medical ZnO for nursery pigs over recent years (DANMAP, 2019). In light of the coming ZnO ban, there is currently a quest for alternatives, and as there may be lessons to learn from the mode of action of this compound, we will briefly touch upon this topic here. The recommended dietary level of Zn for weaner pigs is 100 mg/kg; however, medicinal levels (up to 2500–3000 mg Zn/kg diet) have been used to ameliorate and/or prevent PWD development during the first two weeks post-weaning. This practice was initially stimulated by a report (Poulsen, 1989) showing that 3000 mg Zn as ZnO resulted in a 60% reduction of diarrhoea incidence when compared with lower dietary Zn levels (addition of 0, 100, 200 or 1000 ppm Zn as ZnO). Subsequently, another study showed that a supplement of 2500 ppm Zn as ZnO for two weeks post-weaning (at 28 days of age) reduced the incidence of non-specific PWD by up to 50% and had a growth-promoting effect (Poulsen, 1995). These observations have been supported by more recent reports (Cho et al., 2015, Kaevska et al., 2016, Stensland et al., 2015, Trckova et al., 2015). A large Danish trial with nursery pigs (approximately 3200 animals), in a herd suffering from *E. coli*-mediated PWD, also demonstrated a reduction in the number of diarrhoea treatments by supplementation of 2500 ppm Zn as ZnO in combination with organic acids (Johansen et al., 2007).

The exact mode of action of feeding medicinal levels of ZnO in relation to pig physiology has not been fully elucidated. However, zinc is involved in various biological processes. The effect has been connected to improved piglet feed intake (Broom et al., 2003) and it has further been suggested that Zn as well as Cu additives may prevent the development of physiological Zn and Cu deficiency (Carlson, 2003, Carlson et al., 1999, Hill et al., 2001, Poulsen, 1995). Moreover, studies have found changes in some pancreatic enzymes and hormonal status (Hedemann et al., 2006, Li et al., 2006).

The reported effects of ZnO on the gastrointestinal microbiota point towards the involvement of non-physiological mechanisms. Thus, Højberg et al. (2005) observed a reduced bacterial activity in the gut digesta of pigs receiving 2500 mg/kg Zn as ZnO compared with that in animals receiving 100 mg/kg, most likely reflecting a reduced bacterial load in the gastrointestinal tract of the former. Data like these indicate a major influence of ZnO dose on the gastrointestinal tract microbiota, and show that, besides a potential promotion of feed intake, high dietary ZnO doses may render more energy available for the host animal by a general suppression of the commensal gut microbiota. This has also been suggested as one of the working mechanisms behind the effect of antibiotic growth promoters (Collier et al., 2003, Gaskins et al., 2002). It should be emphasised that it might not only be a question of energy availability for the host animal. Commensal bacteria, like certain lactic acid bacteria, as well as potential pathogens such as *C. perfringens*, may utilise essential feed components, such as amino acids, as well as impair lipid digestibility by hydrolysing (deconjugating) bile salts (Knarreborg et al., 2002a, Knarreborg et al., 2002c).

A decrease in the total number of anaerobes and lactobacilli and an increase in the number of coliforms have been observed in ileum samples from animals fed a high ZnO-amended diet (Højberg et al., 2005, Jensen, 1987). Thus, direct inhibition of potential pathogens, as represented by enterobacteria/coliforms, has typically not been reported for post-weaning piglets receiving high ZnO levels (Jensen, 1987, Højberg et al., 2005, Broom et al., 2006, Vahjen et al., 2010), even though ZnO has been shown to reduce piglet susceptibility to *E. coli* infections (Mores et al., 1998). However, high doses of dietary ZnO have been shown to support a higher diversity of enterobacteria/coliforms in weaned piglets (Katouli et al., 1999, Vahjen et al., 2011) and more complex modes of action may apply, where the impact of dietary ZnO on the diversity of the coliform population may prevent the blooming of, for example, specific pathogenic ETEC strains. Indeed, *in vitro* studies have shown that members of different bacterial taxa, commensals as well as pathogens, show a wide range of ZnO sensitivity (Liedtke and Vahjen, 2012). Thus, the *in vivo* antimicrobial effects of ZnO may be governed by a more selective growth inhibition, that is, the presence of ZnO-sensitive and -resistant/tolerant species or strains within the various microbial genera.

A more recent study by Vahjen et al. (2016) demonstrated that the level of specific *E. coli* toxin genes was decreased by a high ZnO dose (3000 ppm). This study also demonstrated a more detailed time lapse, where *E. coli* were inhibited from 3–4 days post-weaning with a concomitant increase in the faecal ZnO level, demonstrating that the piglets had started to eat. However, the *E. coli* numbers started to increase again 8–14 days post-weaning. This scenario may indicate an adaptation to ZnO by the *E. coli* population. It remains to be answered, however, how the growth-promoting and diarrhoea-reducing effects of excess dietary ZnO are exerted. It seems that a high dietary ZnO level has a positive impact on the stability and diversity of the piglet gastrointestinal microbiota, contributing to an increased colonisation resistance against pathogens, and thereby resistance to diarrhoeal infections, eventually promoting piglet growth performance (Katouli et al., 1999, Vahjen et al., 2011, Pieper et al., 2012b, 2020).

7 Conclusion and future trends in research

The routine use of antibiotics at sub-therapeutic concentrations has been banned in the EU since 1 January 2006 due to the increasing prevalence of antibiotic resistance in pigs. High levels of ZnO are frequently used as a feed additive for pigs to improve gut health and promote growth, and are suggested as an alternative to antimicrobial growth promoters. However, due to the risk of developing antimicrobial-resistant bacteria, this use will be phased out by 2022 in the EU, and therefore alternatives to both antibiotics and high levels of ZnO to control ETEC infections and PWD in post-weaning piglets are of great importance. Future studies focusing on the key components in the delicate balance between dysbiosis/eubiosis and the interaction with nutrition will enable us to develop solutions to enhance gut function to prevent PWD.

8 Where to look for further information

The authors have previously provided information on how to improve gut function in pigs via dietary factors, see for instance Canibe (2019) and Canibe et al. (2003 and 2005). More recently, Lauridsen has in collaboration with both scientific organisations and private industries provided an overview on the role of fatty acids and vitamins on gut function in pigs (Lauridsen, 2020, Lauridsen et al., 2021). However, there are many scientific papers coming up these years on how to improve gut function in pigs via other dietary- and management factors, including knowledge on alternatives to both antibiotics and high levels of ZnO to control ETEC infections and PWD in post-weaning piglets. Upcoming international conferences, such as ZeroZnSummit (ZeroZincSummit2022 (svineproduktion.dk)) arranged by SEGES in Denmark, and Kaessler Nutrition Forum (https://www.kaesler.de/en/kaeslerer-forum-2022/research-submit) in Germany.

- Lauridsen, C. 2020. Effects of dietary fatty acids on gut health and function of pigs pre- and post-weaning. *Journal of Animal Science* 98(4), skaa086.
- Lauridsen, C., Matte, J. J., Lessard, M., Celi, P. and Litta, G. 2021. Role of vitamins for gastro-intestinal functionality and health of pigs. *Animal Feed Sci. Technol.* 273, 114823.

9 References

Abdelli, N., Pérez, J. F., Vilarrasa, E., Cabeza Luna, I., Melo-Duran, D., D'Angelo, M. and Solà-Oriol, D. 2020. Targeted-release organic acids and essential oils improve performance and digestive function in broilers under a necrotic enteritis challenge. *Animals (Basel)* 10(2), 259 (30 pages).

Alexopoulos, C., Georgoulakis, I. E., Tzivara, A., Kritas, S. K., Siochu, A. and Kyriakis, S. C. 2004. Field evaluation of the efficacy of a probiotic containing *Bacillus licheniformis* and *Bacillus subtilis* spores, on the health status and performance of sows and their litters. *Journal of Animal Physiology and Animal Nutrition* 88(11-12), 381–392.

Arguello, H., Estelle, J., Zaldivar-Lopez, S., Jimenez-Marin, Á, Carvajal, A., Lopez-Bascon, M. A., Crispie, F., O'sullivan, O., Cotter, P. D., Priego-Capote, F., Morera, L. and Garrido, J. J. 2018. Early *Salmonella* Typhimurium infection in pigs disrupts microbiome composition and functionality principally at the ileum mucosa. *Scientific Reports* 8(1), 7788 (12 pages).

Bakkali, F., Averbeck, S., Averbeck, D. and Idaomar, M. 2008. Biological effects of essential oils - a review. *Food and Chemical Toxicology: An International Journal Published for the British Industrial Biological Research Association* 46(2), 446–475.

Bao, W. B., Ye, L., Pan, Z. Y., Zhu, J., Du, Z. D., Zhu, G. Q., Huang, X. G. and Wu, S. L. 2012. The effect of mutation at M307 in FUT1 gene on susceptibility of *Escherichia coli* F18 and gene expression in Sutai piglets. *Molecular Biology Reports* 39(3), 3131 –3136.

Barba-Vidal, E., Roll, V. F. B., Castillejos, L., Guerra-Ordaz, A. A., Manteca, X., Mallo, J. J. and Martin-Orue, S. M. 2017. Response to a *Salmonella* Typhimurium challenge in piglets supplemented with protected sodium butyrate or *Bacillus licheniformis*: effects on performance, intestinal health and behavior,2. *Translational Animal Science* 1(2), 186-200.

Barman, M., Unold, D., Shifley, K., Amir, E., Hung, K., Bos, N. and Salzman, N. 2008. Enteric salmonellosis disrupts the microbial ecology of the murine gastrointestinal tract. *Infection and Immunity* 76(3), 907-915.

Beal, J. D., Niven, S. J., Campbell, A. and Brooks, P. H. 2002. The effect of temperature on the growth and persistence of *Salmonella* in fermented liquid pig feed. *International Journal of Food Microbiology* 79(1-2), 99-104.

Bearson, S. M. D., Allen, H. K., Bearson, B. L., Looft, T., Brunelle, B. W., Kich, J. D., Tuggle, C. K., Bayles, D. O., Alt, D., Levine, U. Y. and Stanton, T. B. 2013. Profiling the gastrointestinal microbiota in response to *Salmonella*: low versus high *Salmonella* shedding in the natural porcine host. *Infection, Genetics and Evolution: Journal of Molecular Epidemiology and Evolutionary Genetics in Infectious Diseases* 16, 330-340.

Bennett, K. W. and Eley, A. 1993. Fusobacteria - new taxonomy and related diseases. *Journal of Medical Microbiology* 39(4), 246-254.

Bian, G. R., Ma, S., Zhu, Z. G., Su, Y., Zoetendal, E. G., Mackie, R., Liu, J. H., Mu, C. L., Huang, R. H., Smidt, H. and Zhu, W. Y. 2016. Age, introduction of solid feed and weaning are more important determinants of gut bacterial succession in piglets than breed and nursing mother as revealed by a reciprocal cross-fostering model. *Environmental Microbiology* 18(5), 1566-1577.

Bien, J., Palagani, V. and Bozko, P. 2013. The intestinal microbiota dysbiosis and *Clostridium difficile* infection: is there a relationship with inflammatory bowel disease? *Therapeutic Advances in Gastroenterology* 6(1), 53-68.

Bikker, P., Dirkzwager, A., Fledderus, J., Trevisi, P., Le Huerou-Luron, I., Lalles, J. P. and Awati, A. 2006. The effect of dietary protein and fermentable carbohydrates levels on growth performance and intestinal characteristics in newly weaned piglets. *Journal of Animal Science* 84(12), 3337-3345.

Bindels, L. B., Delzenne, N. M., Cani, P. D. and Walter, J. 2015. Towards a more comprehensive concept for prebiotics. *Nature Reviews. Gastroenterology and Hepatology* 12(5), 303-310.

Boesen, H. T., Jensen, T. K., Schmidt, A. S., Jensen, B. B., Jensen, S. M. and Moller, K. 2004. The influence of diet on *Lawsonia intracellularis* colonization in pigs upon experimental challenge. *Veterinary Microbiology* 103(1-2), 35-45.

Bosi, P., Sarli, G., Casini, L., De Filippi, S., Trevisi, P., Mazzoni, M. and Merialdi, G. 2007. The influence of fat protection of calcium formate on growth and intestinal defence in *Escherichia coli* K88-challenged weanling pigs. *Animal Feed Science and Technology* 139(3-4), 170-185.

Bouyahya, A., Abrini, J., Dakka, N. and Bakri, Y. 2019. Essential oils of Origanum compactum increase membrane permeability, disturb cell membrane integrity, and suppress quorum-sensing phenotype in bacteria. *Journal of Pharmaceutical Analysis* 9(5), 301-311.

Britton, R. A. and Young, V. B. 2012. Interaction between the intestinal microbiota and host in Clostridium difficile colonization resistance. *Trends in Microbiology* 20(7), 313-319.

Brooks, P. H. 2008. Fermented liquid feed for pigs. *CAB Reviews: Perspectives in Agriculture, Veterinary Science, Nutrition and Natural Resources* 3(73), 1-18.

Broom, L. J., Miller, H. M., Kerr, K. G. and Knapp, J. S. 2006. Effects of zinc oxide and *Enterococcus faecium* SF68 dietary supplementation on the performance, intestinal microbiota and immune status of weaned piglets. *Research in Veterinary Science* 80(1), 45-54.

Broom, L. J., Miller, H. M., Kerr, K. G. and Toplis, P. 2003. Removal of both zinc oxide and avilamycin from the post-weaning piglet diet: consequences for performance through to slaughter. *Animal Science* 77(1), 79-84.

Brown, D. R. and Price, L. D. 2008. Catecholamines and sympathomimetic drugs decrease early *Salmonella* Typhimurium uptake into porcine Peyer's patches. *FEMS Immunology and Medical Microbiology* 52(1), 29-35.

Bruininx, E. M., Binnendijk, G. P., Van Der Peet-Schwering, C. M. C., Schrama, J. W., Den Hartog, L. A., Everts, H. and Beynen, A. C. 2002. Effect of creep feed consumption on individual feed intake characteristics and performance of group-housed weanling pigs. *Journal of Animal Science* 80(6), 1413-1418.

Bruininx, E. M. A. M., Schellingerhout, A. B., Binnendijk, G. P., Van Der Peet-Schwering, C. M. Cvd, Schrama, J. W., Den Hartog, L. A., Everts, H. and Beynen, A. C. 2004. Individually assessed creep food consumption by suckled piglets: influence on post-weaning food intake characteristics and indicators of gut structure and hind-gut fermentation. *Animal Science* 78(1), 67-75.

Brunberg, E. I., Rodenburg, T. B., Rydhmer, L., Kjaer, J. B., Jensen, P. and Keeling, L. J. 2016. Omnivores going astray: a review and new synthesis of abnormal behavior in pigs and laying hens. *Frontiers in Veterinary Science* 3, 57.

Burrin, D. and Stoll, B. 2003. Intestinal nutrient requirements in weanling pigs. In: Pluske, J. R., Verstegen, M. W. A. and Le Dividich, J. (Eds) *Weaning the Pig: Concepts and Consequences*, pp. 301-335. Wageningen, The Netherlands: Academic Press Publishers.

Calder, P. C., Albers, R., Antoine, J. M., Blum, S., Bourdet-Sicard, R., Ferns, G. A., Folkerts, G., Friedmann, P. S., Frost, G. S., Guarner, F., Løvik, M., Macfarlane, S., Meyer, P. D., M'rabet, L., Serafini, M., Van Eden, W., Van Loo, J., Vas Dias, W., Vidry, S., Winklhofer-Roob, B. M. and Zhao, J. 2009. Inflammatory disease processes and interactions with nutrition. *British Journal of Nutrition* 101(Suppl. 1), S1-45.

Callegari, M. A., Novais, A. K., Oliveira, E. R., Dias, C. P., Schmoller, D. L., Pereira Junior, M., Nagi, J. G., Alves, J. B. and Silva, C. Ad 2016. Microencapsulated acids associated with essential oils and acid salts for piglets in the nursery phase. *Semina: Ciencias Agrarias* 37(4), 2193-2207.

Callesen, J., Halas, D., Thorup, F., Knudsen, K. E. B., Kim, J. C., Mullan, B. P., Hampson, D. J., Wilson, R. H. and Pluske, J. R. 2007. The effects of weaning age, diet composition, and categorisation of creep feed intake by piglets on diarrhoea and performance after weaning. *Livestock Science* 108(1-3), 120-123.

Canibe, N. 2019. Effects of diet on the gastrointestinal microbial ecosystem and gut gene expression in young pigs. *Animal Nutrition Conference of Canada*, May 15-16. Niagara Falls, Ontario: Sheraton on the Falls, 128-147.

Canibe, N., Hojberg, O., Hojsgaard, S. and Jensen, B. B. 2005. Feed physical form and formic acid addition to the feed affect the gastrointestinal ecology and growth performance of growing pigs. *Journal of Animal Science* 83(6), 1287-1302.

Canibe, N. and Jensen, B. B. 2003. Fermented and nonfermented liquid feed to growing pigs: effect on aspects of gastrointestinal ecology and growth performance. *Journal of Animal Science* 81(8), 2019-2031.

Canibe, N. and Jensen, B. B. 2012. Fermented liquid feed-microbial and nutritional aspects and impact on enteric diseases in pigs. *Animal Feed Science and Technology* 173(1-2), 17-40.

Canibe, N., Steien, S. H., Overland, M. and Jensen, B. B. 2001. Effect of K-diformate in starter diets on acidity, microbiota, and the amount of organic acids in the digestive tract of piglets, and on gastric alterations. *Journal of Animal Science* 79(8), 2123-2133.

Carlson, D. 2003. *The Physiological Role of Dietary Zinc and Copper in Weaned Piglets, with Emphasis on Zinc and Intestinal Mucosal Function.* Copenhagen, Denmark: The Royal Veterinary and Agricultural University.

Carlson, M. S., Hill, G. M. and Link, J. E. 1999. Early- and traditionally weaned nursery pigs benefit from phase-feeding pharmacological concentrations of zinc oxide: effect on metallothionein and mineral concentrations. *Journal of Animal Science* 77(5), 1199-1207.

Carstensen, L., Ersboll, A. K., Jensen, K. H. and Nielsen, J. P. 2005. Escherichia coli post-weaning diarrhoea occurrence in piglets with monitored exposure to creep feed. *Veterinary Microbiology* 110(1-2), 113-123.

Caruso, R., Lo, B. C. and Núñez, G. 2020. Host-microbiota interactions in inflammatory bowel disease. *Nature Reviews. Immunology* 20(7), 411-426.

Castillo, M., Martín-Orúe, S. M., Roca, M., Manzanilla, E. G., Badiola, I., Perez, J. F. and Gasa, J. 2006. The response of gastrointestinal microbiota to avilamycin, butyrate, and plant extracts in early-weaned pigs1,2. *Journal of Animal Science* 84(10), 2725-2734.

Celiberto, L. S., Pinto, R. A., Rossi, E. A., Vallance, B. A. and Cavallini, D. C. U. 2018. Isolation and characterization of potentially probiotic bacterial strains from mice: proof of concept for personalized probiotics. *Nutrients* 10(11), 1684 (21 pages).

Chang, J. Y., Antonopoulos, D. A., Kalra, A., Tonelli, A., Khalife, W. T., Schmidt, T. M. and Young, V. B. 2008. Decreased diversity of the fecal microbiome in recurrent *Clostridium difficile*-associated diarrhea. *Journal of Infectious Diseases* 197(3), 435-438.

Chapman, C. M. C., Gibson, G. R. and Rowland, I. 2011. Health benefits of probiotics: are mixtures more effective than single strains? *European Journal of Nutrition* 50(1), 1-17.

Chen, L. M., Xu, Y. S., Chen, X. Y., Fang, C., Zhao, L. P. and Chen, F. 2017. The maturing development of gut microbiota in commercial piglets during the weaning transition. *Frontiers in Microbiology* 8, 1688.

Cherrington, C. A., Hinton, M., Mead, G. C. and Chopra, I. 1991. Organic acids: chemistry, antibacterial activity and practical applications. *Advances in Microbial Physiology* 32, 87-108.

Cho, J. H., Upadhaya, S. D. and Kim, I. H. 2015. Effects of dietary supplementation of modified zinc oxide on growth performance, nutrient digestibility, blood profiles, fecal microbial shedding and fecal score in weanling pigs. *Animal Science Journal* 86(6), 617-623.

Choi, J., Wang, L., Ammeter, E., Lahaye, L., Liu, S., Nyachoti, M. and Yang, C. B. 2020. Evaluation of lipid matrix microencapsulation for intestinal delivery of thymol in weaned pigs. *Translational Animal Science* 4(1), 411-422.

Chowdhury, S. R., King, D. E., Willing, B. P., Band, M. R., Beever, J. E., Lane, A. B., Loor, J. J., Marini, J. C., Rund, L. A., Schook, L. B., Van Kessel, A. G. and Gaskins, H. R. 2007. Transcriptome profiling of the small intestinal epithelium in germfree versus conventional piglets. *BMC Genomics* 8, 215.

Ciesinski, L., Guenther, S., Pieper, R., Kalisch, M., Bednorz, C. and Wieler, L. H. 2018. High dietary zinc feeding promotes persistence of multi-resistant *E. coli* in the swine gut. *PLoS ONE* 13(1), e0191660.

Clarke, J. M., Topping, D. L., Bird, A. R., Young, G. P. and Cobiac, L. 2008. Effects of high-amylose maize starch and butyrylated high-amylose maize starch on azoxymethane-induced intestinal cancer in rats. *Carcinogenesis* 29(11), 2190-2194.

Collier, C. T., Smiricky-Tjardes, M. R., Albin, D. M., Wubben, J. E., Gabert, V. M., Deplancke, B., Bane, D., Anderson, D. B. and Gaskins, H. R. 2003. Molecular ecological analysis of porcine ileal microbiota responses to antimicrobial growth promoters. *Journal of Animal Science* 81(12), 3035-3045.

Coop, R. L. and Kyriazakis, I. 1999. Nutrition-parasite interaction. *Veterinary Parasitology* 84(3-4), 187-204.

Costa, M. O., Chaban, B., Harding, J. C. S. and Hill, J. E. 2014. Characterization of the Fecal microbiota of Pigs before and after Inoculation with "*Brachyspira hampsonii*". *PLoS ONE* 9(8), e106399.

Dai, Z. L., Wu, G. Y. and Zhu, W. Y. 2011. Amino acid metabolism in intestinal bacteria: links between gut ecology and host health. *Frontiers in Bioscience* 16, 1768-1786.

DANMAP 2019. Use of Antimicrobial Agents and Occurrence of Antimicrobial Resistance in Bacteria from Food Animals, Food and Humans in Denmark. ISSN 1600-2032.

Davis, M. E., Sears, S. C., Apple, J. K., Maxwell, C. V. and Johnson, Z. B. 2006. Effect of weaning age and commingling after the nursery phase of pigs in a wean-to-finish facility on growth, and humoral and behavioral indicators of well-being1,2. *Journal of Animal Science* 84(3), 743-756.

De Graaf, F. K. and Mooi, F. R. 1987. The fimbrial adhesins of *Escherichia coli*. In: Rose, A. H. and Tempest, D. W. (Eds) *Advances in Microbial Physiology*. Academic Press, New York, 28: pp. 65-143.

Diether, N. E. and Willing, B. P. 2019. Microbial fermentation of dietary protein: an important factor in diet-microbe-host interaction. *Microorganisms* 7(1) pp. 1-14 (open access).

Donabedian, H. 2006. Nutritional therapy and infectious diseases: a two-edged sword. *Nutrition Journal* 5, 21.

Dou, S., Gadonna-Widehem, P., Rome, V., Hamoudi, D., Rhazi, L., Lakhal, L., Larcher, T., Bahi-Jaber, N., Pinon-Quintana, A., Guyonvarch, A., Huerou-Luron, I. L. E. and Abdennebi-Najar, L. 2017. Characterisation of early-life fecal microbiota in susceptible and healthy pigs to post-weaning diarrhoea. *PLoS ONE* 12(1), e0169851.

Drumo, R., Pesciaroli, M., Ruggeri, J., Tarantino, M., Chirullo, B., Pistoia, C., Petrucci, P., Martinelli, N., Moscati, L., Manuali, E., Pavone, S., Picciolini, M., Ammendola, S., Gabai, G., Battistoni, A., Pezzotti, G., Alborali, G. L., Napolioni, V., Pasquali, P. and Magistrali, C. F. 2016. *Salmonella enterica* serovar Typhimurium exploits inflammation to modify swine intestinal microbiota. *Frontiers in Cellular and Infection Microbiology* 5, 106.

Dubreuil, J. D. 2017. Enterotoxigenic *Escherichia coli* and probiotics in swine: what the bleep do we know? *Bioscience of Microbiota, Food and Health* 36(3), 75-90.

Fairbrother, J. M., Nadeau, E. and Gyles, C. L. 2005. *Escherichia coli* in postweaning diarrhea in pigs: an update on bacterial types, pathogenesis, and prevention strategies. *Animal Health Research Reviews* 6(1), 17-39.

Fan, P. X., Liu, P., Song, P. X., Chen, X. Y. and Ma, X. 2017. Moderate dietary protein restriction alters the composition of gut microbiota and improves ileal barrier function in adult pig model. *Scientific Reports* 7, 43412.

Faust, K. and Raes, J. 2012. Microbial interactions: from networks to models. *Nature Reviews. Microbiology* 10(8), 538-550.

Ferrara, F., Tedin, L., Pieper, R., Meyer, W. and Zentek, J. 2017. Influence of medium-chain fatty acids and short-chain organic acids on jejunal morphology and intra-epithelial immune cells in weaned piglets. *Journal of Animal Physiology and Animal Nutrition* 101(3), 531-540.

Frese, S. A., Parker, K., Calvert, C. C. and Mills, D. A. 2015. Diet shapes the gut microbiome of pigs during nursing and weaning. *Microbiome* 3, 28.

Frydendahl, K. 2002. Prevalence of serogroups and virulence genes in *Escherichia coli* associated with postweaning diarrhoea and edema disease in pigs and a comparison of diagnostic approaches. *Veterinary Microbiology* 85(2), 169-182.

Gaskins, H. R., Collier, C. T. and Anderson, D. B. 2002. Antibiotics as growth promotants: mode of action. *Animal Biotechnology* 13(1), 29-42.

Gerritsen, R., Van Der Aar, P. and Molist, F. 2012. Insoluble nonstarch polysaccharides in diets for weaned piglets. *Journal of Animal Science* 90(Suppl. 4), 318-320.

Gibson, G. R., Hutkins, R., Sanders, M. E., Prescott, S. L., Reimer, R. A., Salminen, S. J., Scott, K., Stanton, C., Swanson, K. S., Cani, P. D., Verbeke, K. and Reid, G. 2017. Expert consensus document: the International Scientific Association for Probiotics and Prebiotics (ISAPP) consensus statement on the definition and scope of prebiotics. Nature Reviews Gastroenterology & Hepatology 14, 491-502. doi: 10.1038/nrgastro.2017.75.

Gibson, G. R., Scott, K. P., Rastall, R. A., Tuohy, K. M., Hotchkiss, A., Dubert-Ferrandon, A., Gareau, M., Murphy, E. F., Saulnier, D., Loh, G., Macfarlane , S., Delzenne, N., Ringel, Y., Kozianowski, G., Dickmann, R., Lenoir-Wijnkook, I., Walker, C. and Buddington, R. 2010. Dietary prebiotics: current status and new definition. *Food Science and Technology Bulletin: Functional Foods* 7(1), 1-19.

Gilbert, M. S., Ijssennagger, N., Kies, A. K. and Van Mil, S. W. C. 2018. Protein fermentation in the gut; implications for intestinal dysfunction in humans, pigs, and poultry. *American Journal of Physiology. Gastrointestinal and Liver Physiology* 315(2), G159-G170.

Goetstouwers, T., Van Poucke, M., Coppieters, W., Nguyen, V. U., Melkebeek, V., Coddens, A., Van Steendam, K., Deforce, D., Cox, E. and Peelman, L. J. 2014. Refined candidate region for F4ab/ac enterotoxigenic *Escherichia coli* susceptibility situated proximal to MUC13 in Pigs. *PLoS ONE* 9(8), e105013.

Gresse, R., Chaucheyras-Durand, F., Fleury, M. A., Van De Wiele, T., Forano, E. and Blanquet-Diot, S. 2017. Gut microbiota dysbiosis in postweaning piglets: understanding the keys to health. *Trends in Microbiology* 25(10), 851-873.

Grilli, E., Tugnoli, B., Passey, J. L., Stahl, C. H., Piva, A. and Moeser, A. J. 2015. Impact of dietary organic acids and botanicals on intestinal integrity and inflammation in weaned pigs. *BMC Veterinary Research* 11, 96.

Guerra-Ordaz, A. A., Gonzalez-Ortiz, G., La Ragione, R. M., Woodward, M. J., Collins, J. W., Perez, J. F. and Martin-Orue, S. M. 2014. Lactulose and Lactobacillus plantarum, a potential complementary synbiotic to control postweaning colibacillosis in piglets. *Applied and Environmental Microbiology* 80(16), 4879-4886.

Guevarra, R. B., Hong, S. H., Cho, J. H., Kim, B. R., Shin, J., Lee, J. H., Kang, B. N., Kim, Y. H., Wattanaphansak, S., Isaacson, R. E., Song, M. and Kim, H. B. 2018. The dynamics of the piglet gut microbiome during the weaning transition in association with health and nutrition. *Journal of Animal Science and Biotechnology* 9, 54.

Guevarra, R. B., Lee, J. H., Lee, S. H., Seok, M. J., Kim, D. W., Kang, B. N., Johnson, T. J., Isaacson, R. E. and Kim, H. B. 2019. Piglet gut microbial shifts early in life: causes and effects. *Journal of Animal Science and Biotechnology* 10, 1.

Guilloteau, P., Martin, L., Eeckhaut, V., Ducatelle, R., Zabielski, R. and Van Immerseel, F. 2010. From the gut to the peripheral tissues: the multiple effects of butyrate. *Nutrition Research Reviews* 23(2), 366-384.

Halas, D., Hansen, C. F., Hampson, D. J., Mullan, B. P., Wilson, R. H. and Pluske, J. R. 2009. Effect of dietary supplementation with inulin and/or benzoic acid on the incidence and severity of post-weaning diarrhoea in weaner pigs after experimental challenge with enterotoxigenic Escherichia coli. *Archives of Animal Nutrition* 63(4), 267-280.

Hedemann, M. S., Jensen, B. B. and Poulsen, H. D. 2006. Influence of dietary zinc and copper on digestive enzyme activity and intestinal morphology in weaned pigs. *Journal of Animal Science* 84(12), 3310-3320.

Helander, I. M., Alakomi, H.-L., Latva-Kala, K., Mattila-Sandholm, T., Pol, I., Smid, E. J., Gorris, L. G. M. and Von Wright, A. 1998. Characterization of the action of selected essential oil components on Gram-negative bacteria. *Journal of Agricultural and Food Chemistry* 46(9), 3590-3595.

Heo, J. M., Kim, J. C., Hansen, C. F., Mullan, B. P., Hampson, D. J. and Pluske, J. R. 2008. Effects of feeding low protein diets to piglets on plasma urea nitrogen, faecal ammonia nitrogen, the incidence of diarrhoea and performance after weaning. *Archives of Animal Nutrition* 62(5), 343-358.

Heo, J. M., Kim, J. C., Hansen, C. F., Mullan, B. P., Hampson, D. J. and Pluske, J. R. 2009. Feeding a diet with decreased protein content reduces indices of protein fermentation and the incidence of postweaning diarrhea in weaned pigs challenged with an enterotoxigenic strain of Escherichia coli. *Journal of Animal Science* 87(9), 2833-2843.

Heo, J. M., Opapeju, F. O., Pluske, J. R., Kim, J. C., Hampson, D. J. and Nyachoti, C. M. 2013. Gastrointestinal health and function in weaned pigs: a review of feeding strategies to control post-weaning diarrhoea without using in-feed antimicrobial compounds. *Journal of Animal Physiology and Animal Nutrition* 97(2), 207-237.

Hermann-Bank, M. L., Skovgaard, K., Stockmarr, A., Strube, M. L., Larsen, N., Kongsted, H., Ingerslev, H. C., Molbak, L. and Boye, M. 2015. Characterization of the bacterial gut microbiota of piglets suffering from new neonatal porcine diarrhoea. *BMC Veterinary Research* 11, 139.

Hermes, R. G., Molist, F., Ywazaki, M., Nofrarias, M., De Segura, A. G., Gasa, J. and Perez, J. F. 2009. Effect of dietary level of protein and fiber on the productive performance and health status of piglets. *Journal of Animal Science* 87(11), 3569-3577.

Hill, C., Guarner, F., Reid, G., Gibson, G. R., Merenstein, D. J., Pot, B., Morelli, L., Canani, R. B., Flint, H. J., Salminen, S., Calder, P. C. and Sanders, M. E. 2014. Expert consensus document. The International Scientific Association for Probiotics and Prebiotics consensus statement on the scope and appropriate use of the term probiotic. *Nature Reviews. Gastroenterology and Hepatology* 11(8), 506-514.

Hill, G. M., Mahan, D. C., Carter, S. D., Cromwell, G. L., Ewan, R. C., Harrold, R. L., Lewis, A. J., Miller, P. S., Shurson, G. C., Veum, T. L. and NCR-42 Committee on Swine Nutrition 2001. Effect of pharmacological concentrations of zinc oxide with or without the inclusion of an antibacterial agent on nursery pig performance. *Journal of Animal Science* 79(4), 934-941.

Hodgson, K. R. and Barton, M. D. 2009. Treatment and control of enterotoxigenic *Escherichia coli* infections in pigs. *CAB Reviews: Perspectives in Agriculture, Veterinary Science, Nutrition and Natural Resources* 4(44), 1-16.

Holman, D. B., Brunelle, B. W., Trachsel, J. and Allen, H. K. 2017. Meta-analysis to define a core microbiota in the swine gut. *mSystems* 2(3), e00004-17.

Hooks, K. B. and O'Malley, M. A. 2017. Dysbiosis and its discontents. *mBio* 8(5):e01492-17. https://doi.org/10.1128/mBio.01492-17..

Hooper, L. V., Wong, M. H., Thelin, A., Hansson, L., Falk, P. G. and Gordon, J. I. 2001. Molecular analysis of commensal host-microbial relationships in the intestine. *Science* 291(5505), 881-884.

Hopwood, D. E., Pethick, D. W., Pluske, J. R. and Hampson, D. J. 2004. Addition of pearl barley to a rice-based diet for newly weaned piglets increases the viscosity of the intestinal contents, reduces starch digestibility and exacerbates post-weaning colibacillosis. *British Journal of Nutrition* 92(3), 419-427.

Htoo, J. K., Araiza, B. A., Sauer, W. C., Rademacher, M., Zhang, Y., Cervantes, M. and Zijlstra, R. T. 2007. Effect of dietary protein content on ileal amino acid digestibility, growth performance, and formation of microbial metabolites in ileal and cecal digesta of early-weaned pigs. *Journal of Animal Science* 85(12), 3303-3312.

Huang, Y. L., Chassard, C., Hausmann, M., Von Itzstein, M. and Hennet, T. 2015. Sialic acid catabolism drives intestinal inflammation and microbial dysbiosis in mice. *Nature Communications* 6, 8141.

Hutkins, R. W., Krumbeck, J. A., Bindels, L. B., Cani, P. D., Fahey, G., Goh, Y. J., Hamaker, B., Martens, E. C., Mills, D. A., Rastal, R. A., Vaughan, E. and Sanders, M. E. 2016. Prebiotics: why definitions matter. *Current Opinion in Biotechnology* 37, 1-7.

Højberg, O., Canibe, N., Poulsen, H. D., Hedemann, M. S. and Jensen, B. B. 2005. Influence of dietary zinc oxide and copper sulfate on the gastrointestinal ecosystem in newly weaned piglets. *Applied and Environmental Microbiology* 71(5), 2267-2277.

Jakobsen, G. V., Jensen, B. B., Knudsen, K. E. B. and Canibe, N. 2015a. Impact of fermentation and addition of non-starch polysaccharide-degrading enzymes on microbial population and on digestibility of dried distillers grains with solubles in pigs. *Livestock Science* 178, 216-227.

Jakobsen, G. V., Jensen, B. B., Knudsen, K. E. B. and Canibe, N. 2015b. Improving the nutritional value of rapeseed cake and wheat dried distillers grains with solubles by addition of enzymes during liquid fermentation. *Animal Feed Science and Technology* 208, 198-213.

Janczyk, P., Pieper, R., Smidt, H. and Souffrant, W. B. 2007. Changes in the diversity of pig ileal lactobacilli around weaning determined by means of 16S rRNA gene amplification and denaturing gradient gel electrophoresis. *FEMS Microbiology Ecology* 61(1), 132-140.

Jensen, A. N., Dalsgaard, A., Baggesen, D. L. and Nielsen, E. M. 2006. The occurrence and characterization of *Campylobacter jejuni* and *C. coli* in organic pigs and their outdoor environment. *Veterinary Microbiology* 116(1-3), 96-105.

Jensen, B. B. 1987. Tarmfloraen, Zinkoxid og colidiarr, hos svin (Intestinal microflora, zinc oxide and coli enteritis in pigs). *Landbonyt* 41(August), 5-10.

Jimenez, M. J., Berrios, R., Stelzhammer, S. and Bracarense, A. P. F. R. L. 2020. Ingestion of organic acids and cinnamaldehyde improves tissue homeostasis of piglets exposed to enterotoxic *Escherichia coli* (ETEC). *Journal of Animal Science* 98(2), 1–11.

Johansen, M., Jørgensen, L. and Schultz, M. S. 2007. Effekt af zink og organiske Syrer på diarréer i smågriseperioden. Report No 778. Danish Pig Production.

Jørgensen, C. B., Cirera, S., Anderson, S. I., Archibald, A. L., Raudsepp, T., Chowdhary, B., Edfors-Lilja, I., Andersson, L. and Fredholm, M. 2003. Linkage and comparative mapping of the locus controlling susceptibility towards *E. coli* F4ab/ac diarrhoea in pigs. *Cytogenetic and Genome Research* 102(1–4), 157–162.

Kaevska, M., Lorencova, A., Videnska, P., Sedlar, K., Provaznik, I. and Trckova, M. 2016. Effect of sodium humate and zinc oxide used in prophylaxis of post-weaning diarrhoea on faecal microbiota composition in weaned piglets. *Veterinarni Medicina* 61(6), 328–336.

Katouli, M., Melin, L., Jensen-Waern, M., Wallgren, P. and Möllby, R. 1999. The effect of zinc oxide supplementation on the stability of the intestinal flora with special reference to composition of coliforms in weaned pigs. *Journal of Applied Microbiology* 87(4), 564–573.

Ke, S. L., Fang, S. M., He, M. Z., Huang, X. C., Yang, H., Yang, B., Chen, C. Y. and Huang, L. S. 2019. Age-based dynamic changes of phylogenetic composition and interaction networks of health pig gut microbiome feeding in a uniformed condition. *BMC Veterinary Research* 15(1), 172.

Kim, H. B., Borewicz, K., White, B. A., Singer, R. S., Sreevatsan, S., Tu, Z. J. and Isaacson, R. E. 2011a. Longitudinal investigation of the age-related bacterial diversity in the feces of commercial pigs. *Veterinary Microbiology* 153(1–2), 124–133.

Kim, J. C., Heo, J. M., Mullan, B. P. and Pluske, J. R. 2011b. Efficacy of a reduced protein diet on clinical expression of post-weaning diarrhoea and life-time performance after experimental challenge with an enterotoxigenic strain of *Escherichia coli*. *Animal Feed Science and Technology* 170(3–4), 222–230.

Kirchgessner, M., Gedek, B., Wiehler, S., Bott, A., Eidelsburger, U. and Roth, F. X. 1992. Influence of formic-acid, calcium formate and sodium hydrogen carbonate on the microflora in different segments of the gastrointestinal-tract .10. Investigations about the nutritive efficacy of organic-acids in the rearing of piglets. *Journal of Animal Physiology and Animal Nutrition-Zeitschrift fur Tierphysiologie Tierernahrung und Futtermittelkunde* 68, 73–81.

Kluge, H., Broz, J. and Eder, K. 2006. Effect of benzoic acid on growth performance, nutrient digestibility, nitrogen balance, gastrointestinal microflora and parameters of microbial metabolism in piglets. *Journal of Animal Physiology and Animal Nutrition* 90(7–8), 316–324.

Knarreborg, A., Engberg, R. M., Jensen, S. K. and Jensen, B. B. 2002a. Quantitative determination of bile salt hydrolase activity in bacteria isolated from the small intestine of chickens. *Applied and Environmental Microbiology* 68(12), 6425–6428.

Knarreborg, A., Miquel, N., Granli, T. and Jensen, B. B. 2002b. Establishment and application of an in vitro methodology to study the effects of organic acids on coliform and lactic acid bacteria in the proximal part of the gastrointestinal tract of piglets. *Animal Feed Science and Technology* 99(1–4), 131–140.

Knarreborg, A., Simon, M. A., Engberg, R. M., Jensen, B. B. and Tannock, G. W. 2002c. Effects of dietary fat source and subtherapeutic levels of antibiotic on the bacterial community in the ileum of broiler chickens at various ages. *Applied and Environmental Microbiology* 68(12), 5918-5924.

Knudsen, K. E. B., Laerke, H. N., Ingerslev, A. K., Hedemann, M. S., Nielsen, T. S. and Theil, P. K. 2016. Carbohydrates in pig nutrition - recent advances. *Journal of Animal Science* 94, 1-12.

Koropatkin, N. M., Cameron, E. A. and Martens, E. C. 2012. How glycan metabolism shapes the human gut microbiota. *Nature Reviews. Microbiology* 10(5), 323-335.

Kort, R. 2014. Personalized therapy with probiotics from the host by TripleA. *Trends in Biotechnology* 32(6), 291-293.

Kraimi, N., Dawkins, M., Gebhardt-Henrich, S. G., Velge, P., Rychlik, I., Volf, J., Creach, P., Smith, A., Colles, F. and Leterrier, C. 2019. Influence of the microbiota-gut-brain axis on behavior and welfare in farm animals: a review. *Physiology and Behavior* 210, 112658.

Kritas, S. K., Marubashi, T., Filioussis, G., Petridou, E., Christodoulopoulos, G., Burriel, A. R., Tzivara, A., Theodoridis, A. and Piskorikova, M. 2015. Reproductive performance of sows was improved by administration of a sporing bacillary probiotic (*Bacillus subtilis* C-3102). *Journal of Animal Science* 93(1), 405-413.

Kuang, Y., Wang, Y., Zhang, Y., Song, Y., Zhang, X., Lin, Y., Che, L., Xu, S., Wu, D., Xue, B. and Fang, Z. 2015. Effects of dietary combinations of organic acids and medium chain fatty acids as a replacement of zinc oxide on growth, digestibility and immunity of weaned pigs. *Animal Feed Science and Technology* 208, 145-157.

Kuller, W. I., Van Beers-Schreurs, H. M. G., Soede, N. M., Langendijk, P., Taverne, M. A. M., Kemp, B. and Verheijden, J. H. M. 2007. Creep feed intake during lactation enhances net absorption in the small intestine after weaning. *Livestock Science* 108(1-3), 99-101.

Lalles, J. P., Boudry, G., Favier, C., Le Floc'h, N., Lurona, I., Montagne, L., Oswald, I. P., Pie, S., Piel, C. and Seve, B. 2004. Gut function and dysfunction in young pigs: physiology. *Animal Research* 53(4), 301-316.

Lauridsen, C. 2019. From oxidative stress to inflammation: redox balance and immune system. *Poultry Science* 98(10), 4240-4246.

Lee, Y. K. 2018. Personalized probiotics based on phenotypes and dietary habit: a critical evaluation. *Journal of Probiotics and Health* 06(2), 1-4.

Letellier, A., Messier, S., Lessard, L. and Quessy, S. 2000. Assessment of various treatments to reduce carriage of *Salmonella* in swine. *Canadian Journal of Veterinary Research-Revue Canadienne de Recherche Veterinaire* 64(1), 27-31.

Lewis, B. B., Buffie, C. G., Carter, R. A., Leiner, I., Toussaint, N. C., Miller, L. C., Gobourne, A., Ling, L. L. and Pamer, E. G. 2015. Loss of microbiota-mediated colonization resistance to *Clostridium difficile* infection With oral vancomycin compared With metronidazole. *Journal of Infectious Diseases* 212(10), 1656-1665.

Li, S., Zheng, J., Deng, K., Chen, L., Zhao, X. L. L., Jiang, X. M., Fang, Z. F., Che, L., Xu, S. Y., Feng, B., Li, J., Lin, Y., Wu, Y. Y., Han, Y. M. and Wu, D. 2018a. Supplementation with organic acids showing different effects on growth performance, gut morphology, and microbiota of weaned pigs fed with highly or less digestible diets. *Journal of Animal Science* 96(8), 3302-3318.

Li, Y., Guo, Y., Wen, Z. S., Jiang, X. M., Ma, X. and Han, X. Y. 2018b. Weaning stress perturbs gut microbiome and its metabolic profile in piglets. *Scientific Reports* 8(1), 18068.

Li, X., Yin, J., Li, D., Chen, X., Zang, J. and Zhou, X. 2006. Dietary supplementation with zinc oxide increases IGF-I and IGF-I receptor gene expression in the small intestine of weanling piglets. *Journal of Nutrition* 136(7), 1786-1791.

Li, Z., Li, D., Yi, G., Yin, J. and Sun, P. 2007. Effect of organic acids and antibiotic growth promoters on growth performance, gastrointestinal pH, intestinal microbial populations and immune responses of weaned pigs. *Journal of Animal Science* 85, 644-644.

Li, Z. H., Cai, M., Liu, Y. S., Sun, P. L. and Luo, S. L. 2019. Antibacterial activity and mechanisms of essential oil from *Citrus medica* L. var. sarcodactylis. *Molecules* 24(8), 1577 (10 pages).

Liao, S. F. F. and Nyachoti, M. 2017. Using probiotics to improve swine gut health and nutrient utilization. *Animal Nutrition* 3(4), 331-343.

Liedtke, J. and Vahjen, W. 2012. In vitro antibacterial activity of zinc oxide on a broad range of reference strains of intestinal origin. *Veterinary Microbiology* 160(1-2), 251-255.

Lindecrona, R. H., Jensen, T. K. and Moller, K. 2004. Influence of diet on the experimental infection of pigs with *Brachyspira pilosicoli*. *Veterinary Record* 154(9), 264-267.

Lochmiller, R. L. and Deerenberg, C. 2000. Trade-offs in evolutionary immunology: just what is the cost of immunity? *Oikos* 88(1), 87-98.

Lozupone, C. A., Stombaugh, J. I., Gordon, J. I., Jansson, J. K. and Knight, R. 2012. Diversity, stability and resilience of the human gut microbiota. *Nature* 489(7415), 220-230.

Luise, D., Correa, F., Bosi, P. and Trevisi, P. 2020. A review of the effect of formic acid and its salts on the gastrointestinal microbiota and performance of pigs. *Animals: An Open Access Journal from MDPI* 10(5).

Luise, D., Lauridsen, C., Bosi, P. and Trevisi, P. 2019. Methodology and application of *Escherichia coli* F4 and F18 encoding infection models in post-weaning pigs. *Journal of Animal Science and Biotechnology* 10, 53.

Luise, D., Le Sciellour, M., Buchet, A., Resmond, R., Clement, C., Rossignol, M.-N., Jardet, D., Zemb, O., Belloc, C. and Merlot, E. 2021. The fecal microbiota of piglets during weaning transition and its association with piglet growth across various farm environments. *PLoS ONE* 16, e0250655.

Luise, D., Motta, V., Salvarani, C., Chiappelli, M., Fusco, L., Bertocchi, M., Mazzoni, M., Maiorano, G., Costa, L. N., Van Milgen, J., Bosi, P. and Trevisi, P. 2017. Long-term administration of formic acid to weaners: influence on intestinal microbiota, immunity parameters and growth performance. *Animal Feed Science and Technology* 232, 160-168.

Macfarlane, G. T. and Macfarlane, S. 2012. Bacteria, colonic fermentation, and gastrointestinal health. *Journal of AOAC International* 95(1), 50-60.

Mach, N., Berri, M., Estelle, J., Levenez, F., Lemonnier, G., Denis, C., Leplat, J. J., Chevaleyre, C., Billon, Y., Dore, J., Rogel-Gaillard, C. and Lepage, P. 2015. Early-life establishment of the swine gut microbiome and impact on host phenotypes. *Environmental Microbiology Reports* 7(3), 554-569.

Madec, F., Bridoux, N., Bounaix, S. and Jestin, A. 1998. Measurement of digestive disorders in the piglet at weaning and related risk factors. *Preventive Veterinary Medicine* 35(1), 53-72.

Mallo, J. J., Balfagon, A., Gracia, M. I., Honrubia, P. and Puyalto, M. 2012. Evaluation of different protections of butyric acid aiming for release in the last part of the gastrointestinal tract of piglets. *Journal of Animal Science* 90(Suppl. 4), 227-229.

Mao, L. and Franke, J. 2015. Symbiosis, dysbiosis, and rebiosis-The value of metaproteomics in human microbiome monitoring. *Proteomics* 15(5-6), 1142-1151.

Massacci, F. R., Berri, M., Lemonnier, G., Guettier, E., Blanc, F., Jardet, D., Rossignol, M. N., Mercat, M. J., Doré, J., Lepage, P., Rogel-Gaillard, C. and Estellé, J. 2020. Late weaning is associated with increased microbial diversity and *Faecalibacterium prausnitzii* abundance in the fecal microbiota of piglets. *Animal Microbiome* 2(1), 2.

McDonald, D. E., Pethick, D. W., Pluske, J. R. and Hampson, D. J. 1999. Adverse effects of soluble non-starch polysaccharide (guar gum) on piglet growth and experimental colibacillosis immediately after weaning. *Research in Veterinary Science* 67(3), 245-250.

McLamb, B. L., Gibson, A. J., Overman, E. L., Stahl, C. and Moeser, A. J. 2013. Early weaning stress in pigs impairs innate mucosal immune responses to enterotoxigenic *E. coli* challenge and exacerbates intestinal injury and clinical disease. *PLoS ONE* 8(4), e59838.

Medland, J. E., Pohl, C. S., Edwards, L. L., Frandsen, S., Bagley, K., Li, Y. and Moeser, A. J. 2016. Early life adversity in piglets induces long-term upregulation of the enteric cholinergic nervous system and heightened, sex-specific secretomotor neuron responses. *Neurogastroenterology and Motility: The Official Journal of the European Gastrointestinal Motility Society* 28(9), 1317-1329.

Melkebeek, V., Goddeeris, B. M. and Cox, E. 2013. ETEC vaccination in pigs. *Veterinary Immunology and Immunopathology* 152(1-2), 37-42.

Metzler-Zebeli, B. U., Trevisi, P., Prates, J. A. M., Tanghe, S., Bosi, P., Canibe, N., Montagne, L., Freire, J. and Zebeli, Q. 2017. Assessing the effect of dietary inulin supplementation on gastrointestinal fermentation, digestibility and growth in pigs: a meta-analysis. *Animal Feed Science and Technology* 233, 120-132.

Mikkelsen, L. L., Jakobsen, M. and Jensen, B. B. 2003. Effects of dietary oligosaccharides on microbial diversity and fructo-oligosaccharide degrading bacteria in faeces of piglets post-weaning. *Animal Feed Science and Technology* 109(1-4), 133-150.

Mikkelsen, L. L. and Jensen, B. B. 2004. Effect of fructo-oligosaccharides and transgalacto-oligosaccharides on microbial populations and microbial activity in the gastrointestinal tract of piglets post-weaning. *Animal Feed Science and Technology* 117(1-2), 107-119.

Missotten, J. A. M., Michiels, J., Degroote, J. and De Smet, S. 2015. Fermented liquid feed for pigs: an ancient technique for the future. *Journal of Animal Science and Biotechnology* 6(1), 4.

Moeser, A. J., Pohl, C. S. and Rajput, M. 2017. Weaning stress and gastrointestinal barrier development: implications for lifelong gut health in pigs. *Animal Nutrition* 3(4), 313-321.

Molist, F., Van Oostrum, M., Perez, J. F., Mateos, G. G., Nyachoti, C. M. and Van Der Aar, P. J. 2014. Relevance of functional properties of dietary fibre in diets for weanling pigs. *Animal Feed Science and Technology* 189, 1-10.

Mores, N., Christani, J., Piffer, I. A., Barioni, Jr. W. and Lima, G. M. M. 1998. Efeito do oxido de zinco no controle da diarreia pos-desmame em leitoes infectados experimentalmente com *Escherichia coli* (Effects of zinc oxide on postweaning diarrhea control in pigs experimentally infected with *E. coli*). *Arquivo Brasileiro de Medicina veterinaria e Zootecnia* 50, 513-523.

Mori, K., Ito, T., Miyamoto, H., Ozawa, M., Wada, S., Kumagai, Y., Matsumoto, J., Naito, R., Nakamura, S., Kodama, H. and Kurihara, Y. 2011. Oral administration of multispecies microbial supplements to sows influences the composition of gut microbiota and fecal organic acids in their post-weaned piglets. *Journal of Bioscience and Bioengineering* 112(2), 145–150.

Mosca, A., Leclerc, M. and Hugot, J. P. 2016. Gut microbiota diversity and human diseases: should we reintroduce key predators in our ecosystem? *Frontiers in Microbiology* 7, 455.

Motta, V., Luise, D., Bosi, P. and Trevisi, P. 2019. Faecal microbiota shift during weaning transition in piglets and evaluation of AO blood types as shaping factor for the bacterial community profile. *PLoS ONE* 14(5), e0217001.

Mulder, I. E., Schmidt, B., Stokes, C. R., Lewis, M., Bailey, M., Aminov, R. I., Prosser, J. I., Gill, B. P., Pluske, J. R., Mayer, C. D., Musk, C. C. and Kelly, D. 2009. Environmentally-acquired bacteria influence microbial diversity and natural innate immune responses at gut surfaces. *BMC Biology* 7, 79.

Nagy, B. and Fekete, P. Z. 1999. Enterotoxigenic *Escherichia coli* (ETEC) in farm animals. *Veterinary Research* 30(2–3), 259–284.

Nair, M. S., Eucker, T., Martinson, B., Neubauer, A., Victoria, J., Nicholson, B. and Pieters, M. 2019. Influence of pig gut microbiota on *Mycoplasma hyopneumoniae* susceptibility. *Veterinary Research* 50, 86 (13 pages).

Namkung, H., Li J. Gong, M., Yu, H., Cottrill, M. and De Lange, C. F. M. 2004. Impact of feeding blends of organic acids and herbal extracts on growth performance, gut microbiota and digestive function in newly weaned pigs. *Canadian Journal of Animal Science* 84(4), 697–704.

Nielsen, T. S., Bendiks, Z., Thomsen, B., Wright, M. E., Theil, P. K., Scherer, B. L. and Marco, M. L. 2019. High-amylose maize, potato, and butyrylated starch modulate large intestinal fermentation, microbial composition, and oncogenic miRNA expression in rats fed a high-protein meat diet. *International Journal of Molecular Sciences* 20(9), 2137 (19 pages).

Nyachoti, C. M., Omogbenigun, F. O., Rademacher, M. and Blank, G. 2006. Performance responses and indicators of gastrointestinal health in early-weaned pigs fed low-protein amino acid-supplemented diets. *Journal of Animal Science* 84(1), 125–134.

Omonijo, F. A., Ni, L., Gong, J., Wang, Q., Lahaye, L. and Yang, C. 2018. Essential oils as alternatives to antibiotics in swine production. *Animal Nutrition* 4(2), 126–136.

Opapeju, F. O., Krause, D. O., Payne, R. L., Rademacher, M. and Nyachoti, C. M. 2009. Effect of dietary protein level on growth performance, indicators of enteric health, and gastrointestinal microbial ecology of weaned pigs induced with postweaning colibacillosis. *Journal of Animal Science* 87(8), 2635–2643.

Pajarillo, E. A. B., Chae, J., P. Balolong, M., Bum Kim, H. and Kang, D. 2014. Assessment of fecal bacterial diversity among healthy piglets during the weaning transition. *Journal of General and Applied Microbiology* 60(4), 140–146.

Papatsiros, V. G., Tassis, P. D., Tzika, E. D., Papaioannou, D. S., Petridou, E., Alexopoulos, C. and Kyriakis, S. C. 2011. Effect of benzoic acid and combination of benzoic acid with a probiotic containing *Bacillus cereus* var. toyoi in weaned pig nutrition. *Polish Journal of Veterinary Sciences* 14(1), 117–125.

Parker, A., Lawson, M. A. E., Vaux, L. and Pin, C. 2018. Host-microbe interaction in the gastrointestinal tract. *Environmental Microbiology* 20(7), 2337–2353.

Partanen, K., Siljander-Rasi, H., Pentikainen, J., Pelkonen, S. and Fossi, M. 2007. Effects of weaning age and formic acid-based feed additives on pigs from weaning to slaughter. *Archives of Animal Nutrition* 61(5), 336–356.

Patil, Y., Gooneratne, R. and Ju, X. H. 2020. Interactions between host and gut microbiota in domestic pigs: a review. *Gut Microbes* 11(3), 310-334.

Petersen, C. and Round, J. L. 2014. Defining dysbiosis and its influence on host immunity and disease. *Cellular Microbiology* 16(7), 1024-1033.

Pieper, R., Boudry, C., Bindelle, J., Vahjen, W. and Zentek, J. 2014. Interaction between dietary protein content and the source of carbohydrates along the gastrointestinal tract of weaned piglets. *Archives of Animal Nutrition* 68(4), 263-280.

Pieper, R., Dadi, T. H., Pieper, L., Vahjen, W., Franke, A., Reinert, K. and Zentek, J. 2020. Concentration and chemical form of dietary zinc shape the porcine colon microbiome, its functional capacity and antibiotic resistance gene repertoire. *The ISME Journal* 14(11), 2783-2793.

Pieper, R., Kroger, S., Richter, J. F., Wang, J., Martin, L., Bindelle, J., Htoo, J. K., Von Smolinski, D., Vahjen, W., Zentek, J. and Van Kessel, A. G. 2012a. Fermentable fiber ameliorates fermentable protein-induced changes in microbial ecology, but not the mucosal response, in the colon of piglets. *Journal of Nutrition* 142(4), 661-667.

Pieper, R., Vahjen, W., Neumann, K., Van Kessel, A. G. and Zentek, J. 2012b. Dose-dependent effects of dietary zinc oxide on bacterial communities and metabolic profiles in the ileum of weaned pigs. *Journal of Animal Physiology and Animal Nutrition* 96(5), 825-833.

Pieper, R., Tudela, C. V., Taciak, M., Bindelle, J., Perez, J. F. and Zentek, J. 2016. Health relevance of intestinal protein fermentation in young pigs. *Animal Health Research Reviews* 17(2), 137-147.

Pluske, J. R. 2013. Feed- and feed additives-related aspects of gut health and development in weanling pigs. *Journal of Animal Science and Biotechnology* 4(1), 1.

Pluske, J. R. 2016. Invited review: aspects of gastrointestinal tract growth and maturation in the pre- and postweaning period of pigs. *Journal of Animal Science* 94(Suppl_3), 399-411.

Pohl, C. S., Medland, J. E., Mackey, E., Edwards, L. L., Bagley, K. D., Dewilde, M. P., Williams, K. J. and Moeser, A. J. 2017. Early weaning stress induces chronic functional diarrhea, intestinal barrier defects, and increased mast cell activity in a porcine model of early life adversity. *Neurogastroenterology and Motility: The Official Journal of the European Gastrointestinal Motility Society* 29(11), e13118.

Portrait, V., Cottenceau, G. and Pons, A. M. 2000. A *Fusobacterium mortiferum* strain produces a bacteriocin-like substance(s) inhibiting *Salmonella enteritidis*. *Letters in Applied Microbiology* 31(2), 115-117.

Poulsen, A.-S. R., Luise, D., Curtasu, M. V., Sugiharto, S., Canibe, N., Trevisi, P. and Lauridsen, C. 2018a. Effects of alpha-(1,2)-fucosyltransferase genotype variants on plasma metabolome, immune responses and gastrointestinal bacterial enumeration of pigs pre- and post-weaning. *PLoS ONE*, 13(8), e0202970.

Poulsen, A.-S. R., De Jonge, N., Nielsen, J. L., Hojberg, O., Lauridsen, C., Cutting, S. M. and Canibe, N. 2018b. Impact of *Bacillus* spp. spores and gentamicin on the gastrointestinal microbiota of suckling and newly weaned piglets. *PLoS ONE* 13(11), e0207382.

Poulsen, H. D. 1989. Zinc oxide for weaned pigs. Proceedings of the 40th Annual Meeting in the European Association for Animal Production 1989, 8-10, (abstr. P5.14).

Poulsen, H. D. 1995. Zinc oxide for weanling piglets. *Acta Agriculturae Scandinavica, Section A - Animal Science* 45(3), 159-167.

Pu, J. N., Chen, D. W., Tian, G., He, J., Zheng, P., Mao, X. B., Yu, J., Huang, Z. Q., Zhu, L., Luo, J. Q., Luo, Y. H. and Yu, B. 2018. Protective Effects of benzoic acid, *Bacillus coagulans*, and Oregano Oil on intestinal Injury Caused by enterotoxigenic *Escherichia coli* in Weaned Piglets. *BioMed Research International* 2018, 1829632.

Rattigan, R., Sweeney, T., Vigors, S., Rajauria, G. and O'doherty, J. V. 2020. Effects of reducing dietary crude protein concentration and supplementation with laminarin or zinc oxide on the faecal scores and colonic microbiota in newly weaned pigs. *Journal of Animal Physiology and Animal Nutrition* 104(5), 1471-1483.

Rauw, W. M. 2012. Immune response from a resource allocation perspective. *Frontiers in Genetics* 3, 267.

Reid, G., Gaudier, E., Guarner, F., Huffnagle, G. B., Macklaim, J. M., Munoz, A. M., Martini, M., Ringel-Kulka, T., Sartor, B., Unal, R., Verbeke, K., Walter, J. and International Scientific Association for Probiotics and Prebiotics 2010. Responders and non-responders to probiotic interventions How can we improve the odds? *Gut Microbes* 1(3), 200-204.

Rist, V. T. S., Weiss, E., Eklund, M. and Mosenthin, R. 2013. Impact of dietary protein on microbiota composition and activity in the gastrointestinal tract of piglets in relation to gut health: a review. *Animal: An International Journal of Animal Bioscience* 7(7), 1067-1078.

Roca, M., Nofrarias, M., Majo, N., De Rozas, A. M. P., Segales, J., Castillo, M., Martin-Orue, S. M., Espinal, A., Pujols, J. and Badiola, I. 2014. Changes in bacterial population of gastrointestinal tract of weaned pigs fed with different additives. *BioMed Research International* 2014, 269402.

Rodríguez-Sorrento, A., Castillejos, L., López-Colom, P., Cifuentes-Orjuela, G., Rodríguez-Palmero, M., Moreno-Muñoz, J. A. and Martín-Orúe, S. M. 2020. Effects of *Bifidobacterium longum* Subsp. infantis CECT 7210 and *Lactobacillus rhamnosus* HN001, combined or not With oligofructose-enriched inulin, on weaned pigs orally challenged with *Salmonella* Typhimurium. *Frontiers in Microbiology* 11, 2012.

Roselli, M., Pieper, R., Rogel-Gaillard, C., De Vries, H., Bailey, M., Smidt, H. and Lauridsen, C. 2017. Immunomodulating effects of probiotics for microbiota modulation, gut health and disease resistance in pigs. *Animal Feed Science and Technology* 233, 104-119.

Round, J. L. and Mazmanian, S. K. 2009. The gut microbiota shapes intestinal immune responses during health and disease. *Nature Reviews. Immunology* 9(5), 313-323.

Russell, J. B. 1992. Another explanation for the toxicity of fermentation acids at low pH - anion accumulation versus uncoupling. *Journal of Applied Bacteriology* 73(5), 363-370.

Russell, P. J., Geary, T. M., Brooks, P. H. and Campbell, A. 1996. Performance, water use and effluent output of weaner pigs fed ad libitum with either dry pellets or liquid feed and the role of microbial activity in the liquid feed. *Journal of the Science of Food and Agriculture* 72(1), 8-16.

Salehi, B., Zucca, P., Erdogan Orhan, I. E., Azzini, E., Adetunji, C. O., Mohammed, S. A., Banerjee, S. K., Sharopov, F., Rigano, D., Sharifi-Rad, J., Armstrong, L., Martorell, M., Sureda, A., Martins, N. L., Selamoglu, Z. and Ahmad, Z. 2019. Allicin and health: a comprehensive review. *Trends in Food Science and Technology* 86, 502-516.

Sanders, M. E., Benson, A., Lebeer, S., Merenstein, D. J. and Klaenhammer, T. R. 2018. Shared mechanisms among probiotic taxa: implications for general probiotic claims. *Current Opinion in Biotechnology* 49, 207-216.

Schokker, D., Fledderus, J., Jansen, R., Vastenhouw, S. A., de Bree, F. M., Smits, M. A. and Jansman, A. 2018. Supplementation of fructooligosaccharides to suckling piglets affects intestinal microbiota colonization and immune development. *Journal of Animal Science* 96, 2139-2153. doi: 10.1093/jas/sky110.

Singh, V., Yeoh, B. S., Xiao, X., Kumar, M., Bachman, M., Borregaard, N., Joe, B. and Vijay-Kumar, M. 2015. Interplay between enterobactin, myeloperoxidase and lipocalin 2 regulates *E. coli* survival in the inflamed gut. *Nature Communications* 6, 7113.

Slifierz, M. J., Friendship, R. M. and Weese, J. S. 2015. Longitudinal study of the early-life fecal and nasal microbiotas of the domestic pig. *BMC Microbiology* 15(1), 184.

Starke, I. C., Pieper, R., Neumann, K., Zentek, J. and Vahjen, W. 2013. Individual responses of mother sows to a probiotic *Enterococcus faecium* strain lead to different microbiota composition in their offspring. *Beneficial Microbes* 4(4), 345-356.

Stecher, B. 2015. The roles of inflammation, nutrient availability and the commensal microbiota in enteric pathogen infection. *Microbiology Spectrum* 3(3), MBP-0008-2014.

Stensland, I., Kim, J. C., Bowring, B., Collins, A. M., Mansfield, J. P. and Pluske, J. R. 2015. A comparison of diets supplemented with a feed additive containing organic acids, cinnamaldehyde and a permeabilizing complex, or zinc oxide, on post-weaning diarrhoea, selected bacterial populations, blood measures and performance in weaned pigs experimentally infected with enterotoxigenic *E. coli. Animals: An Open Access Journal from MDPI* 5(4), 1147-1168.

Stojanović-Radić, Z., Matejić, J. and Radulović, N. 2013. Garlic essential oil: biological activities and further potential. In: Govil, J. N. and Bhattacharaya, S. (Eds) Recent Progrees in Medicinal Plants: Essential Oils, Studium Press, ISBN: 9781626993297 pp. 289-328.

Stokes, C. R., Bailey, M., Haverson, K., Harris, C., Jones, P., Inman, C., Pie, S., Oswald, I. P., Williams, B. A., Akkermans, A. D. L., Sowa, E., Rothkotter, H. J. and Miller, B. G. 2004. Postnatal development of intestinal immune system in piglets: implications for the process of weaning. *Animal Research* 53(4), 325-334.

Suiryanrayna, M. V. and Ramana, J. V. 2015. A review of the effects of dietary organic acids fed to swine. *Journal of Animal Science and Biotechnology* 6, 45.

Sulabo, R. C., Jacela, J. Y., Tokach, M. D., Dritz, S. S., Goodband, R. D., Derouchey, J. M. and Nelssen, J. L. 2010. Effects of lactation feed intake and creep feeding on sow and piglet performance. *Journal of Animal Science* 88(9), 3145-3153.

Sun, W., Sun, J., Li, M., Xu, Q., Zhang, X., Tang, Z., Chen, J., Zhen, J. and Sun, Z. 2020. The effects of dietary sodium butyrate supplementation on the growth performance, carcass traits and intestinal microbiota of growing-finishing pigs. *Journal of Applied Microbiology* 128(6), 1613-1623.

Swanson, K. S., Gibson, G. R., Hutkins, R., Reimer, R. A., Reid, G., Verbeke, K., Scott, K. P., Holscher, H. D., Azad, M. B., Delzenne, N. M. and Sanders, M. E. 2020. The International Scientific Association for Probiotics and Prebiotics (ISAPP) consensus statement on the definition and scope of Synbiotics. *Nature Reviews. Gastroenterology and Hepatology* 17(11), 687-701.

Sylvia, K. E. and Demas, G. E. 2018. A gut feeling: microbiome-brain-immune interactions modulate social and affective behaviors. *Hormones and Behavior* 99, 41-49.

Taras, D., Vahjen, W., Macha, M. and Simon, O. 2006. Performance, diarrhea incidence, and occurrence of *Escherichia coli* virulence genes during long-term administration of a probiotic *Enterococcus faecium* strain to sows and piglets. *Journal of Animal Science* 84(3), 608-617.

Taube, V. A., Neu, M. E., Hassan, Y., Verspohl, J., Beyerbach, M. and Kamphues, J. 2009. Effects of dietary additives (potassium diformate/organic acids) as well as influences of grinding intensity (coarse/fine) of diets for weaned piglets experimentally infected with *Salmonella* Derby or *Escherichia coli*. *Journal of Animal Physiology and Animal Nutrition* 93(3), 350-358.

Tiels, P., Verdonck, F., Coddens, A., Goddeeris, B. and Cox, E. 2008. The excretion of F18+E. coli is reduced after oral immunisation of pigs with a FedF and F4 fimbriae conjugate. *Vaccine* 26(17), 2154-2163.

Torrallardona, D., Badiola, I. and Broz, J. 2007. Effects of benzoic acid on performance and ecology of gastrointestinal microbiota in weanling piglets. *Livestock Science* 108(1-3), 210-213.

Tran, T. H. T., Everaert, N. and Bindelle, J. 2018. Review on the effects of potential prebiotics on controlling intestinal enteropathogens *Salmonella* and *Escherichia coli* in pig production. *Journal of Animal Physiology and Animal Nutrition* 102(1), 17-32.

Trckova, M., Lorencova, A., Hazova, K. and Zajacova, Z. S. 2015. Prophylaxis of post-weaning diarrhoea in piglets by zinc oxide and sodium humate. *Veterinarni Medicina* 60, 351-360.

Trombetta, D., Castelli, F., Sarpietro, M. G., Venuti, V., Cristani, M., Daniele, C., Saija, A., Mazzanti, G. and Bisignano, G. 2005. Mechanisms of antibacterial action of three monoterpenes. *Antimicrobial Agents and Chemotherapy* 49(6), 2474-2478.

Tsiloyiannis, V. K., Kyriakis, S. C., Vlemmas, J. and Sarris, K. 2001. The effect of organic acids on the control of post-weaning oedema disease of piglets. *Research in Veterinary Science* 70(3), 281-285.

Turpin, D. L., Langendijk, P., Sharp, C. and Pluske, J. R. 2017. Improving welfare and production in the peri-weaning period: effects of co-mingling and intermittent suckling on the stress response, performance, behaviour, and gastrointestinal tract carbohydrate absorption in young pigs. *Livestock Science* 203, 82-91.

Vahjen, W., Pieper, R. and Zentek, J. 2010. Bar-coded pyrosequencing of 16S rRNA gene amplicons reveals changes in ileal porcine bacterial communities due to high dietary zinc intake. *Applied and Environmental Microbiology* 76(19), 6689-6691.

Vahjen, W., Pieper, R. and Zentek, J. 2011. Increased dietary zinc oxide changes the bacterial core and enterobacterial composition in the ileum of piglets. *Journal of Animal Science* 89(8), 2430-2439.

Vahjen, W., Rom, O. A. and Zentek, J. 2016. Impact of zinc oxide on the immediate postweaning colonization of enterobacteria in pigs. *Journal of Animal Science* 94, 359-376.

Van Noten, N., Degroote, J., Van Liefferinge, E., Taminiau, B., De Smet, S., Desmet, T. and Michiels, J. 2020. Effects of thymol and thymol alpha-D-Glucopyranoside on intestinal function and microbiota of weaned pigs. *Animals: An Open Access Journal from MDPI* 10(2), 21.

Van Winsen, R. L., Lipman, L. J. A., Biesterveld, S., Urlings, B. A. P., Snijders, J. M. A. and Van Knapen, F. 2001a. Mechanism of *Salmonella* reduction in fermented pig feed. *Journal of the Science of Food and Agriculture* 81(3), 342-346.

Van Winsen, R. L., Urlings, B. A. P., Lipman, L. J. A., Snijders, J. M. A., Keuzenkamp, D., Verheijden, J. H. M. and Van Knapen, F. 2001b. Effect of fermented feed on the microbial population of the gastrointestinal tracts of pigs. *Applied and Environmental Microbiology* 67(7), 3071-3076.

Veljovic, K., Dinic, M., Lukic, J., Mihajlovic, S., Tolinacki, M., Zivkovic, M., Begovic, J., Mrvaljevic, I., Golic, N. and Terzic-Vidojevic, A. 2017. Promotion of early gut colonization by probiotic intervention on microbiota diversity in pregnant sows. *Frontiers in Microbiology* 8, 2028.

Verdonck, F., Cox, E. and Goddeeris, B. M. 2004. F4 fimbriae expressed by porcine enterotoxigenic *Escherichia coli*, an example of an eccentric fimbrial system? *Journal of Molecular Microbiology and Biotechnology* 7(4), 155-169.

Verdonck, F., Tiels, P., Van Gog, K., Goddeeris, B. M., Lycke, N., Clements, J. and Cox, E. 2007. Mucosal immunization of piglets with purified F18 fimbriae does not protect against F18+ *Escherichia coli* infection. *Veterinary Immunology and Immunopathology* 120(3-4), 69-79.

Wang, B. H., Yao, M. F., Lv, L. X., Ling, Z. X. and Li, L. J. 2017. The human microbiota in health and disease. *Engineering* 3(1), 71-82.

Wang, Y. Q., Zhou, Y. T., Xiao, X., Zheng, J. and Zhou, H. 2020. Metaproteomics: a strategy to study the taxonomy and functionality of the gut microbiota. *Journal of Proteomics* 219, 103737 (11 pages).

Weber, N., Nielsen, J. P., Jakobsen, A. S., Pedersen, L. L., Hansen, C. F. and Pedersen, K. S. 2015. Occurrence of diarrhoea and intestinal pathogens in non-medicated nursery pigs. *Acta Veterinaria Scandinavica* 57, 64.

Wei, H. K., Xue, H. X., Zhou, Z. X. and Peng, J. 2017. A carvacrol-thymol blend decreased intestinal oxidative stress and influenced selected microbes without changing the messenger RNA levels of tight junction proteins in jejunal mucosa of weaning piglets. *Animal: An International Journal of Animal Bioscience* 11(2), 193-201.

Wellock, I. J., Fortomaris, P. D., Houdijk, J. G. M. and Kyriazakis, I. 2006. The effect of dietary protein supply on the performance and risk of post-weaning enteric disorders in newly weaned pigs. *Animal Science* 82(3), 327-335.

Wellock, I. J., Fortomaris, P. D., Houdijk, J. G. M. and Kyriazakis, I. 2007. Effect of weaning age, protein nutrition and enterotoxigenic *Escherichia coli* challenge on the health of newly weaned piglets. *Livestock Science* 108(1-3), 102-105.

Wellock, I. J., Fortomaris, P. D., Houdijk, J. G. M. and Kyriazakis, I. 2008a. Effects of dietary protein supply, weaning age and experimental enterotoxigenic *Escherichia coli* infection on newly weaned pigs: health. *Animal: An International Journal of Animal Bioscience* 2(6), 834-842.

Wellock, I. J., Fortomaris, P. D., Houdijk, J. G. M., Wiseman, J. and Kyriazakis, I. 2008b. The consequences of non-starch polysaccharide solubility and inclusion level on the health and performance of weaned pigs challenged with enterotoxigenic *Escherichia coli*. *British Journal of Nutrition* 99(3), 520-530.

Wijtten, P. J. A., Meulen, J. V. D. and Verstegen, M. W. A. 2011. Intestinal barrier function and absorption in pigs after weaning: a review. *British Journal of Nutrition* 105(7), 967-981.

Xing, Y. Y., Li, K. N., Xu, Y. Q., Wu, Y. Z., Shi, L. L., Guo, S. W., Yan, S. M., Jin, X. and Shi, B. L. 2020. Effects of galacto-oligosaccharide on growth performance, feacal microbiota, immune response and antioxidant capability in weaned piglets. *Journal of Applied Animal Research* 48(1), 63-69.

Xu, Y. T., Liu, L., Long, S. F., Pan, L. and Piao, X. S. 2018. Effect of organic acids and essential oils on performance, intestinal health and digestive enzyme activities of weaned pigs. *Animal Feed Science and Technology* 235, 110-119.

Yang, C., Chowdhury, M. A., Huo, Y. and Gong, J. 2015. Phytogenic compounds as alternatives to in-feed antibiotics: potentials and challenges in application. *Pathogens* 4(1), 137–156.

Yang, Q. L., Huang, X. Y., Zhao, S. G., Sun, W. Y., Yan, Z. Q., Wang, P. F., Li, S. G., Huang, W. Z., Zhang, S. W., Liu, L. X. and Gun, S. B. 2017. Structure and function of the fecal microbiota in diarrheic neonatal piglets. *Frontiers in Microbiology* 8, 502.

Zeng, M. Y., Inohara, N. and Nuñez, G. 2017. Mechanisms of inflammation-driven bacterial dysbiosis in the gut. *Mucosal Immunology* 10(1), 18–26.

Zhou, L., Fang, L., Sun, Y., Su, Y. and Zhu, W. 2016. Effects of the dietary protein level on the microbial composition and metabolomic profile in the hindgut of the pig. *Anaerobe* 38, 61–69.

Chapter 4

Improving gut function in pigs to prevent pathogen colonization

P. Bosi, D. Luise and P. Trevisi, University of Bologna, Italy

1 Introduction

Intestinal pathogens in livestock cause either clinical or subclinical infections which increase morbidity and (or) mortality, causing economic loss as well as increasing the negative health and environmental impact of animal food products. Optimal functioning of the gastrointestinal tract (GIT) provides a foundation to ensure optimal animal performance at all stages of its life. This is particularly true in pigs which, among domestic species, give birth to the most immature newborn. This is a legacy of life in the wild, where a single litter has minimal contact with others and the long suckling phase ensures the correct development of various gut functions:

- digestive function;
- gastro-intestinal cell line barrier;

http://dx.doi.org/10.19103/AS.2021.0089.16

- gut-associated lymphoid tissue (GALT); and
- gut-associated microbiota.

Increasing genetic progress in breeding animals has resulted in larger litter size. The need to breed numerous animals rapidly in intensive production systems has resulted in the early weaning of pigs, typically between 1 and 4 weeks post-farrowing, depending on individual countries. The intensification of animal production systems also limits the ability of reared animals to express their ancestral natural behavior. In addition, modern pig production chains, based on the multisite system, often involve several transfers between sites, resulting in acute and chronic stress and underestimated effects on the health of pigs. These conditions both inhibit gut development and impose greater pressures on gut function.

Until recently the widespread use of antibiotics has guaranteed sanitary control in all phases of pig production, especially when acute stress events expose pigs to gut inflammation and/or gut dysbiosis, as in weaning or after a transfer to a different production site. However, this use of antibiotics is no longer acceptable in modern production systems. To minimize the problem of antibiotic resistance, the use of antibiotics is now restricted to those classes of antibiotics which are not needed to treat human diseases and limited as much as possible to the treatment of ill animals only. Some countries (e.g. those of the European Union [EU]) have banned additives such as zinc oxide designed to promote growth and prevent post-weaning diarrhoea (PWD) in order to limit the co-selection of antibiotic-resistant microorganisms and to reduce the environmental impact of intensive livestock production.

The attention of experts, technicians and breeders is now to focus on intervention measures that, without compromising production efficiency, improve pig gut functionality from an early age to prevent intestinal colonization of pathogen microorganisms and food-borne pathogens. These measures need to take into account three areas:

- porcine genetics;
- management measures; and
- dietary interventions.

2 The main gut-related pathogens in pigs

This section discusses the following pathogens colonizing the gut: *Escherichia coli*, *Salmonella* spp., *Brachyspira* spp. and *Lawsonia intracellularis*.

2.1 Escherichia coli

Escherichia coli (*E. coli*) is a gram-negative flagellated bacterium, belonging to the Enterobacteriaceae family. Strains of enterotoxigenic *E. coli* (ETEC) are considered the main etiological pathogens associated with neonatal enteric colibacillosis and PWD. These are major problems in pig management, leading to an increase in morbidity and mortality of piglets (Luppi, 2017). It has been estimated that the costs of managing PWD and the losses it causes vary between €40 and €314 per sow (Luppi, 2017).

ETEC has an oro-fecal route of transmission and, once in the intestine, it can adhere to specific receptors in the brush border of the intestinal mucosa, resulting in their infection. ETEC adhesion is mediated by fimbriae and diffuse adherence. ETEC causing neonatal enteric colibacillosis mostly possess the F4, F5, F6 and F41 fimbriae, while post-weaning ETEC mostly possess F4 (45%) and F18 (33.9%) fimbriae (Luppi et al., 2016). ETEC F4 has three serological variants, namely F4ab, F4ad and F4ac (which have been recognized as the most prevalent), whilst ETEC F18 has two antigenic variances, namely F18ab and F18ac (Luppi, 2017). More detailed information on fimbriae receptors can be found in Dubreuil et al. (2016) and Luise et al. (2019a).

ETEC are known to produce heat-labile (LT) enterotoxins and heat-stable (STa, STb and enteroaggregative heat-stable toxin 1 [EAST1]) enterotoxins. Neonatal ETEC mainly produce the heat-stable enterotoxin STa that leads to electrolyte end fluid secretion through a cGMP pathway in the host. Post-weaning ETEC produce both LT and ST enterotoxins; ETEC F18ab can also produce Shiga toxins, causing the disease edema in growing pigs (Nagy and Fekete, 2005). Clinical signs of neonatal enteric colibacillosis include diarrhea in white and yellow colors with a watery to a creamy consistency, while a yellowish-grey watery diarrhea and a reduced appetite leading to sudden death characterize PWD (Luppi, 2017).

2.2 Salmonella *spp.*

Salmonella is a major foodborne pathogen with a significant number of cases and outbreaks every year causing losses in the EU of €3 billion a year (EFSA, 2016). Together with eggs and poultry meat, pork represents the major source of *Salmonella* infection, accounting for 27% of cases across the EU (EFSA, 2009). In addition to the economic impact of salmonellosis on the human population, it is also a major economic disease of pigs, causing millions of dollars in lost income to the pork industry (Roof et al., 1992). *Salmonella* spp. are a common bacteria of pig intestinal microbiota. Among the 2400 different serotypes that have been identified, the two most prevalent in pigs are (Fedorka-Cray and Wray, 2000; Gray et al., 1995):

- *Salmonella enterica* serovar Typhimurium, associated with enterocolitis; and
- *Salmonella enterica* serovar Choleraesuis, associated with enterocolitis and septicemia.

Enterocolitic salmonellosis occurs most frequently in pigs from weaning to 4 months of age (Wilcock and Schwartz, 1992). Salmonellas have an oro-fecal route of transmission and have many potential virulence factors (over 200) (see Kadhim, 2020). *S. enterica* serovar Typhimurium localizes generally along the intestinal tract and has a low tendency to invade the enteric mucosa. *S. enterica* serovar Choleraesuis is preferably localized in the colon or the luminal surface of the ileal Peyer's patches. Clinical signs of enterocolitis are watery and yellow diarrhea without mucus and occasionally blood, anorexia and dehydration, with mortality generally low (Wilcock and Olander, 1977). Septicemia occurs mainly in weaned pigs and growing pigs and occasionally in finishing pigs. Gross lesions of septicemia are visible in several organs including the liver, gall bladder, spleen, lung, stomach and intestine (Wilcock and Schwartz, 1992). The growing recognition of *Salmonella* as a public health problem is associated with the fact that many susceptible pigs remain healthy carriers and can carry *Salmonella* in feces, requiring the development of biosecurity measures and national programs for Salmonella surveillance in pigs (Andres and Davies, 2015).

2.3 Brachyspira *spp.*

Brachyspira is a gram-negative, motile, helically coiled (spiral-shaped), anaerobic bacterium that belongs to the *Brachyspiraceae* family. *B. hyodysenteriae* and *B. pilosicoli* are considered to be the most important pig intestinal pathogens that cause swine dysentery (SD) and porcine intestinal spirochetosis (PIS), respectively. Swine dysentery is a severe mucohemorhagic enteric disease with a worldwide distribution that primarily affects pigs during the growth and finishing periods, leading to major losses caused by mortality and sub-optimal performance (Wills, 2000). The incidence of SD decreased in the 1990s but increased recently (Burrough, 2017); this has been linked in some countries with the withdrawal of carbadox as medication (Hampson et al., 2015). Clinical signs can vary from soft, yellow to gray feces and mild mucous diarrhea to severe hemorrhagic diarrhea that can lead to high levels of mortality (50-90%) (Walczak, 2015). On endemically infected farms, clinical signs often reappear cyclically at 3- to 4-week intervals corresponding to the removal of antimicrobials (Burrough, 2017).

    B. hyodysenteriae is anaerobic but can survive in the feces and the environment. Following ingestion from feces, it survives the acidic environment of the stomach and can reach the large intestine where it invades the mucus and crypts of the mucosa and can penetrate the colonic enterocyte and goblet cells. The epithelial barrier can favor the growth of opportunistic pathogens including the protozoan *Balantidium coli* that causes intestinal damage. For more detailed information on *B. hyodysenteriae*, see Alvarez-Ordóñez et al. (2013), Burrough (2017), Hampson and Burrought (2019).

PIS is associated with loss of condition resulting in a lower rate of growth that increases the time required for the pigs to reach market weight, resulting in a significant increase in production costs (Duhamel, 1998). It occurs in the post-weaning period but also in finishing pigs and occasionally in pregnant sows. PIS has been mainly associated with *B. pilosicoli*, but it can also be due to weakly hemolytic *Brachyspira* species, including *B. intermedia* and *B. murdochii* (Hampson and Burrought, 2019). Its incidence can vary, being low and limited to a small group of animals in some cases whilst being widespread in other herds. The widespread diffusion and reappearance of PIS in a farm have been associated with the presence of different *Brachyspira* species (Oxberry and Hampson, 2003). *B. pilosicoli* is shed intermittently or continuously in the feces and can then survive in feces and in feces mixed with the soil for 210 days (Boye et al., 2001). Once in the large intestine, *B. pilosicoli* cells adhere in the cecal and colonic epithelial cells and particularly the mature apical enterocytes between crypt units (Trott et al., 1996). Some animals can develop a subclinical infection. The main clinical signs are loose and sticky feces in adult pigs and watery to mucoid diarrhea (green or brown) in growers and weaners; in both cases, mortality is rare (Duhamel, 2001).

2.4 Lawsonia intracellularis

L. intracellularis is an obligate intracellular pathogen causing proliferative enteropathy syndromes (PE). PE represents a serious health problem on swine farms all over the world and in Europe. PE prevalence ranges from 20% to 50% in Europe and 75% to 78% in the United States (Chouet et al., 2003; Marsteller et al., 2003). *L. intracellularis* has an oro-fecal route of transmission, infection starting with the entry in the epithelial cells of the crypts through the formation and rupture of cytoplasmic vacuoles (Kroll et al., 2005). *L. intracellularis* replicates in the macrophage, a key factor in the development of its hemorrhagic form (McIntyre et al., 2003). PE can be subclinical or present non-specific clinical signs including grayish-yellow soft stools, presence of undigested food and/ or, more rarely, blood in feces. PE generally leads to weight loss and slower growth.

Based on the clinical signs and lesions, PE has been associated with porcine intestinal adenomatosis (PIA), typical of young growing pigs, proliferative hemorrhagic enteropathy (PHE), more often observed in adult pigs, as well as with less common necrotic enteritis (NE) (McOrist and Gebhart, 1999). The typical signs of PIA include anorexia, diarrhea and persistent poor growth (Lawson and Gebhart, 2000; McOrist and Gebhart, 1999). PHE is characterized by the occasional presence of loose and red watery feces and can lead to death of half of the affected pigs, while the other half recover without a significant difference in body weight (Rowland and Lawson, 1992). More detailed information about the infection, clinical signs, diagnosis and prevalence can be found in Stege et al. (2004), Kroll et al. (2005), Dors et al. (2015) and Van der Heijden et al. (2004).

3 Pig genetics and resistance to disease

The interplay between host genetics, gut microbial composition and host physiological responses has crucial importance for animal health and performance, and it can play a key role in the resistance of pigs to intestinal pathogens. It is well documented that genetic variation influences immune responsiveness in pigs (Zhao et al., 2012; Joling et al., 1993; Edfors-Lilja et al., 1994). It is also well known that the prevalence of enteric and respiratory disease can be inherited (Lundeheim, 1979) and that breed can affect disease heritability and resistance to some specific pathogens (Geraci et al., 2019). However, there is no clear consensus on specific genetic markers and breeding strategies for improving pig genetic resistance to pathogenic disease.

Some genetic markers have been associated with PWD (Coddens et al., 2008; Jørgensen et al., 2003; Ren et al., 2012) and intestinal homeostasis (Bäckhed, 2011; Priori et al., 2016). The susceptibility of pigs to the main pathogens causing PWD, ETEC F4 and ETEC F18 depends on the degree of expression of specific receptors in the small intestine, which allow the pathogen to adhere to the intestinal epithelium, resulting in the consequent development of the disease. Genetic studies have revealed the association between the intestinal expression of specific receptors, susceptibility to ETEC infections and the presence of specific mutations in the porcine genome. There are several mutations located in the Mucin 4 gene (*MUC4*) and the Mucine 13 (*MUC13*) and Mucine 20 (*MUC20*) genes, the transferrin receptor (*TFRC*), tyrosine kinase non-receptor 2 (*ACK1*), and UDP-GlcNAc:betaGal beta-1,3-N-acetylglucosaminyltransferase 5 (*B3GNT5*) gene. These have been associated with susceptibility to ETEC F4 (Jørgensen et al., 2003; Goetstouwers et al., 2014; Ren et al., 2012; Zhang et al., 2008; Ji et al., 2011; Wang et al., 2007; Ouyang et al., 2012). The mutations identified by Liu et al. (2013) on the bactericidal/permeability-increasing protein (*BPI*) gene and by Meijerink et al. (2000) in the Fucosyltransferase 1 (*FUT1*) gene have been associated with susceptibility to ETEC F18. Recent studies investigating the effect of genetic background on gut microbiota found that the specific markers located in the *MUC4* (Jørgensen et al., 2003) and *FUT1* (Meijerink et al., 2000) genes can significantly affect the intestinal and fecal microbial profile of piglets around weaning (Luise et al., 2019b; Poulsen et al., 2018; Massacci et al., 2020), emphasizing the interplay between host genetics, microbiota and immune response.

Together with some SNPs located in the Caveolin-1 gene (*Cav1*) (Liu et al., 2011), the mutation on the *FUT1* gene has also been associated with porcine resistance to *Haemophilus parasuis* (Wang et al., 2012). The *FUT1* gene encodes for alpha (1,2)-fucosyltransferases (FUTs) enzymes that play a key role in the formation of blood group antigens of the AO porcine blood group system (Sako et al., 1990). The FUT gene determines the synthesis of blood group antigens (Galb1,3/ 4(Fuca1,2) GlcNAcb1-R), transferring A fucose to H precursors (Galb1,3/4GlcNAcb1-R). Antigen A formation is due to the N-acetylgalactosaminyltransferase [A3GALNT] encoded by the α 1-3-N- AO gene. Linked to the *FUT1* gene, the AO porcine blood group (AO) has been related to intestinal disease and the modulation of gut microbiota composition in humans (Mäkivuokko et al., 2012) but not in pigs (Motta et al., 2019). However, it has been shown that the A0 blood group genotype can alter the jejunal glycomic binding pattern profile in pigs (Priori et al., 2016).

No specific genetic markers have been found relating to genetic resistance to Salmonella. However, studies have suggested that some specific regions in different porcine chromosomes are associated with the presence of *S. Choleraesuis* in the spleen and liver (Galina-Pantoja et al., 2009). Several candidate SNPs located in haptoglobin (*HP*), neutrophil cytosolic factor 2 (*NCF2*), phosphogluconate dehydrogenase (*PGD*) and Toll-like receptor 4 (*TLR4*) genes are associated with *S. enterica* serovar *Typhimurium* shedding (Uthe et al., 2011; Kich et al., 2014).

In the case of *Lawsonia* spp. and *Streptococcus suis*, no specific genomic regions (or genes and SNPs) related to resistance in pigs have been identified and less information is available. Studies based on gene expression analysis have suggested that some genes may be involved in resistance. Gene encoding for insulin-like growth factor binding protein 3 (*IGFBP-3*) was found to be upregulated in pigs with *L. intracellularis* infection (Jacobson et al., 2011). The gene matrix metalloproteinase 9 (*Mmp9*), toll-like receptor 2 (*Tlr2*), tumor necrosis factor (*Tnf*) and pentraxin 3 (*Ptx3*) have been indicated as candidate resistance gene in mice against *S. suis* infection (Rong et al., 2012).

Compared to other traits, selection for more resistant animals remains challenging. The fact that resistant traits are quantitative and that identification of the infection is not always easy (e.g. *S. suis* infection is frequently asymptomatic in pigs), together with a currently fragmented approach, makes disease resistance breeding still far from practical. However, novel genomics and integration of omics techniques may help to improve breeding for disease resistance, allowing a reduction in antimicrobial use.

4 Management strategies affecting gut functionality and pathogen colonization

Management strategies can affect the gut homeostasis of pigs. They can either promote or reduce susceptibility to intestinal pathogen colonization. Understanding the risk factors and conditions associated with gut mucosa impairment can help farmers to manage risks and avoid detrimental effects on animal health. The neonatal and weaning transition phases are known to be particularly important for establishing proper gut functionality, contributing to preventing pathogen colonization in later life. Among risk factors, heat stress, in particular, can compromise pig gut function.

4.1 Colostrum and milk intake

The period from the first moment after birth to the end of the weaning phase has a major influence on pig development. A key early input in gut development and protection is colostrum. Colostrum provides energy, amino acids, micronutrients and immunoglobulins. It also supplies non-specific defense factors such as lactoferrin (Elliot et al., 1984), lysozyme (Krakowski et al., 2002) and the lactoperoxidase system (Albera and Kankofer, 2009). Sow colostrum also provides an array of growth factors, hormones for enterocytes (Hurley, 2015) and/or local lymphocytes (Cytokines, Hurley, 2015; Osteopontin, Palombo et al., 2018).

Colostrum is also the first main substrate for microbial development. The first microorganisms to be established in the digestive tract include *E. coli* and *Clostridia* spp. (Kubasova et al., 2017), representing the Proteobacteria and Firmicutes phyla (Chen et al., 2018c) and the *Streptococci* (Chen et al., 2018a), transmitted from floors and the suckling milk environment (Chen et al., 2018a,c, Zhang et al., 2018), maternal feces (Kubasova et al., 2017) or the vagina (Zhang et al., 2018).

Colostrum nutrients and defense proteins promote *Fusobacterium* and members of the *Bacteroides* genus (Kubasova et al., 2017; Chen et al., 2018c). The shift from colostrum to milk increases lactose content, promoting the *Lactobacillus* genus (Chen et al., 2018c) as well as members of the *Ruminococcaceae* family (Chen et al., 2018c). The quality of solid feed supplements and individual intake patterns can promote microbes degrading structured carbohydrates, including *Prevotella ruminicola* (Zhang et al., 2016), contributing to microbiota variability.

Oligosaccharide content is quite variable in colostrum (Trevisi et al., 2020) and milk (Cheng et al., 2016). This variability, with specialized sugar complexes based, for example, on sialic acid and fucose, contributes to gut microbiota variation. Fucosylated oligosaccharides can promote the colonization of the gut by bacteria such as the fucose-utilizing *Enterobacteriaceae* strains (Salcedo et al., 2016). The different patterns of oligosaccharides in sow colostrum can explain part of the variability of litter weight gain within 3 days and at weaning but not piglet mortality within 3 days (Trevisi et al., 2020).

Although the composition of colostrum and milk is regarded as important for the development of immunity in piglets, there is no direct evidence of specific protection against pathogen colonization except for immunoglobulins (Igs). Colostrum IgG antibodies are rapidly absorbed into the bloodstream of suckling piglets, almost within the first day of life, provided enough colostrum is consumed (160–170 g/kg body weight, Le Dividich et al., 2005). Any factor (physical inability of the neonate, insufficient number of teats, prolonged farrowing, poor design of the farrowing crates, etc.) limiting this opportunity can impair uptake. IgAs subsequently become the main immunoglobulins in milk, protecting piglets from the aggregation of potential pathogens in the gut lumen. It is possible that, as in mice, the primitive form of immune response of polyreactive IgA provided in milk controls microbial community development of the neonate (Harris et al., 2006). However, specific control of the main intestinal pathogens in neonate pigs relies on the specific acquired immune experience in the gut of the mother due to the entero-mammary link. Specific antibodies in colostrum/milk may not protect against pathogens that the sow did not previously encounter.

However, it is generally accepted that the luminal presence of IgA from milk delays the development of specific immunity in the piglet (Hodgins et al., 1999; Nguyen et al., 2006). In swine, GALT is compartmentalized in: (1) the diffuse system, intraepithelial (IEL) and in the lamina propria, (2) organized compartments: Peyer's patches in jejunum and ileum and mesenteric lymph nodes. After birth, lymphocytes, almost absent at the epithelial surface and the lamina propria of the small intestine, develop rapidly and functionally differentiate under the stimulation of antigens in the gut. This occurs mostly in the first 4–5 weeks and continues to a lesser degree during the juvenile and mature stages (Rothkötter et al., 1999). Peyer's patches begin to organize during the first week of life, reaching a relatively normal architecture at 10–15 days. IgM+ B cells occur in higher numbers but IgA+ B cells develop later (Levast et al., 2014) and a significant antigen repertoire diversification takes place.

After early colostrum intake, milk, though still important, does not determine GALT development. Compared to conventional rearing, artificial feeding of piglets after 1 week of age does not affect the number of Peyer's patches (Prims et al., 2018) or densities of CD8+ and CD4+ cells, the main types of functional lymphocyte (Prims et al., 2016). However artificial rearing reduces the density of M cells related to antigens and ileal Peyer's patches (but not in the tonsils of the soft palate) (Prims et al., 2017). Overall, the modest difference in GALT development can be associated with a temporary variation in the composition of the intestinal microbiome of piglets fed milk formula (less Gram+ microbiome) that disappears by the end of the rearing period (Prims et al., 2016). Conversely, shifting to artificial feeding at 2 days of age reduces lymphoid follicle size and germinal centers in Peyer's patches (Yeruva et al., 2016). It also increases *Streptococcus, Blautia, Citrobacter, Butrycimonas, Parabacteroides* and *Lactococcus* genera and reduces microbial diversity in the colon (Saraf et al., 2017) and the duodenum contents but not in the ileum (Piccolo et al., 2017).

One poorly considered biological component associated with gut maturation in newborn piglets is bacteriophages. The presence of bacteria in the gut induces prophage development in human babies, seen in the presence of virus-like particles (Liang et al., 2020). Lee et al. (2016) added a cocktail of bacteriophages to the creep feed given to suckling pigs which targeted *E. coli* (F4, K99 and F41) and *Clostridium perfringens,* types A and C. This reduced coliform and *Clostridium* spp. fecal shedding in piglets and their mothers.

The evolution of the main enzymes involved in colostrum and milk digestion is well documented (Aumaitre et al., 1995) and this favors the maintenance of villus integrity. In wild animals, access to other sources besides milk increases host enzyme activity and gut commensals. This function is provided by creep feed to suckling piglets. The importance of creep feed to the post-weaning health of piglets depends on the age when supplementation starts, the age at weaning, the quality and the composition of the creep feed, as well as individual intake before and after weaning.

4.2 Pre- and post-weaning feed intake

Achieving a mature, successfully functioning gut requires balancing digestible and absorbable feed for the host with nutrients available for gut commensals. There is a mutualistic relationship between host and microbiota. Host mucins are degraded by bacteria while metabolites produced from microbial fermentation are essential for the host (e.g. volatile fatty acids and vitamins)

As discussed in the previous section, colostrum and milk intake are essential to the proper development of the gut. In confined rearing systems, the limited availability of substrates other than milk results in immature gut mucosa at weaning, especially in digesting complex carbohydrates. Early ingestion of a few grams of solid feed could stimulate the switch of gut mucosa to produce an array of digestive enzymes targeting the digestion of vegetable substrates. Controlled creep feed intake (by gastric intubation) before (6 days) and after (5 days) weaning in pigs weaned at 14 days of age (Kelly et al., 1990b) did not affect mucosal weight, villus morphometry, and mucosal carbohydrase activity, compared with piglets only suckled on milk. Late weaning (20 days of age) allowed the option of studying the duration of creep feed allowance (13 or 6 or

2 days) (Sulabo et al., 2010). A longer duration of creep feeding increased the number of piglets in the litter consuming the supplement, with these subjects eating more in the post-weaning period compared to non-eaters. However, eating more creep feed in the immediate post-weaning was also a predisposing factor for the intestinal proliferation of E. coli O149 in these piglets (Carstensen et al., 2005). This could counteract the positive effect of supplementary feeding during the suckling period, partially preventing the decrease of net absorption of uninfected and ETEC-infected segments of the intestine, usually observed after weaning (Nabuurs et al., 1996). The early presence of ETEC can remain undetected for several days during lactation. At weaning, the prevalence, duration or severity of diarrhea artificially induced by early infection with different ETEC strains was not reduced by creep feed (Kelly et al., 1990a). The different early condition of suckling pigs may explain the different response to creep feeding. For piglets given early supplementary creep feed (week 3), a strong negative correlation was observed between Prevotella and Escherichia in healthy samples, while this correlation was relatively weaker in diarrheic piglets (Yang et al., 2019).

Preventive measures to reduce gut inflammation and/or dysbiosis in the post-weaning period start with preventing the overload of undigested feed in the gut in the first days after birth. The restriction of feeding on the day of weaning is a frequently used practice to prevent diarrhea (Sørensen et al., 2009). However, the controlled prevention of feeding for a day and a half, after feed intake starts, can cause signs of intestinal stress, associated with the intestinal mucosa and villous-crypt atrophy (Lallès and David, 2011). This practice affects energy provision in the first days after weaning in maintaining villus barrier integrity and T lymphocytes (CD4+ to CD8+, Spreeuwenberg et al., 2001). Pigs are kept in groups and individual milk intake and weight may vary with weaning age and the hierarchy established in feeding. This helps explain inconsistencies in some experimental observations, showing limits in individual feed restriction in preventing diarrhea in pigs inoculated or not with E. coli O149 (Sørensen et al., 2009). Group-penned piglets subject to fasting for the first 18 h after weaning (at 7 weeks) are able to compensate rapidly and attain higher body weights 3 weeks after weaning (Millet et al., 2019). Nevertheless, the efficacy of preventive fastening of group-penned pigs in inhibiting pathogen colonization is still unproven, particularly for pigs weaned at early ages (3-4 weeks).

4.3 Heat stress

Other than weaning, environmental conditions play a key role in the physiological homeostasis of pigs. Climate change is increasing areas with warmer temperatures in the spring and summer. Heat stress has a negative effect on pig production in several countries. Growing-finishing pigs and sows are the most susceptible animals to this kind of stress with severe consequences on production performance and health (Ross et al., 2015; Ross et al., 2017). An increase in temperature results in a temporary reduction of voluntary feed intake as an adaptive measure to reduce body temperature. In extreme conditions (between 29°C and 35°C), as observed in the meta-analyses published by de Oliveira et al. (2019), the growth performance of pigs is strongly impaired. Heat stress can also predispose animals to illnesses by impairing gut homeostasis. In order to reduce body core temperature, vasodilation transfers blood flow from splanchnic tissue to the periphery, causing temporary gut hypoxia which compromises mucosa integrity by increasing its permeability.

In pigs of around 60 kg body weight, short exposure to high temperatures reduces intestinal transepithelial resistance, causing damage in the thigh junctions which results in increasing mucosa permeability, a factor that predisposes pigs to bacterial translocation and infection (Pearce et al., 2014). In a study modeling diurnal heat stress in younger pigs (average 21 kg body weight), Gabler et al. (2018) observed an increase of gut mucosa permeability associated with an increase of 150% of serum lipopolysaccharide (LPS) compared with pigs reared in thermoneutral conditions. Increased mucosa permeability can be ascribed to oxidative stress, associated with decreased glutathione peroxidase (GPX) activity and an increased glutathione disulfide (GSSG)-to-glutathione (GSH) ratio (Liu et al., 2016). Measures to contain oxidative stress-induced gut permeability are essential to limit pathogen colonization and translocation, ensuring both animal health and food safety.

5 Dietary strategies to improve gut functionality and prevent pathogen colonization: feed size

The feed can be used to promote defense mechanisms against pathogen colonization. These mechanisms include hydrochloric acid production, promoting the mucin layer of the mucosa, production of antimicrobial peptides and supporting the presence of bile salts.

Feeding a coarsely ground meal feed to pigs changes the physiochemical and microbial properties of feed contents in the stomach, decreasing the survival of pathogens during passage through the stomach (Mikkelsen t al., 2004). The stomach is the first point of entry for bacteria in the GIT. Gastric mucosa harbors a specific microbial community in pigs characterized by *Herbiconiux* and *Brevundimonas*, two genera which include cellulolytic and xylanolytic strains which affect health, including the risk of gastric ulcers (Motta et al., 2017).

The effect of the physical form of diets and pre-processing grinding has been studied in pigs at different ages. In pigs from 7 to 12 weeks of age, fine particles <0.4 mm constitute a higher risk factor for pre-ulcerative epithelial alterations compared to coarse particles. Since the pelleting process increases the prevalence of fine particles, it is best avoided to prevent gastric ulceration (Liesner et al., 2009). These findings have been confirmed in a study by Millet et al. (2012) where crumbled feed was found to impair gastric mucosa integrity and was associated with the increased prevalence of *Helicobacter suis* colonization in the stomach when compared with a coarse meal diet.

The increased ulceration risk from feeding pigs with ground and pelleted feed seems to be due to the increase of chime viscosity rather than to gastric pH (Mößeler et al., 2010). When pigs weighing from 30 kg to 100 kg were fed a coarse diet, Mikkelsen et al. (2004) observed increased concentrations of organic acids and a reduced pH in the stomach, with a potentially positive effect on the mortality of *S. enterica* serovar Typhimurium DT12 (*in vitro* observation). *S. enterica* serovar Typhimurium DT12 adhered less to the ileal mucosa in piglets fed a pelleted diet compared to a mash diet (Hedemann et al., 2005). Mikkelsen et al. (2004) also reported a reduction of coliform bacteria in the small and large intestines when pigs were fed a course diet, associated with the augmentation of butyric acid in the hindgut due to an increase in bacterial fermentation. The ileum of pigs fed a pelleted diet had also more diffuse staining for mucins. This suggests that the thin layer of mucins in the small intestine is capable of upregulating its *Salmonella*-entrapping capability. The mucin-inducing effect of fiber in preventing *Salmonella* infection has recently been confirmed (Wellington et al., 2020). A pelleted diet was found to promote *L. intracellularis* in the ileal microbiota more than a non-pelleted diet (Mølbak et al., 2008). The use of milk whey powder was also associated with a lower *L. intracellularis* caecal content, presumably explained by an increase of acetate and butyrate in the cecum of fattening pigs (Visscher et al., 2018).

6 Dietary strategies to improve gut functionality and prevent pathogen colonization: protein and dietary fiber

It is well known that the ratio and type of protein and fiber play a key role in intestinal health. The quantity of fermentable substrates from diets can affect microbial ecology in the GIT. The promotion of beneficial (lactic acid bacteria) or opportunist/pathogens bacteria determine the formation of metabolites that affect host intestinal health (Jha and Berrocoso, 2016; Aumiller et al., 2015).

6.1 Crude protein

Protein digestion and absorption can be influenced by external factors (gastric ulcers, limited digesta acidification, inflammatory processes) and endogenous factors (reduced feed intake due to stress, feed size, diet viscosity and buffering capacity, antinutritional factors) (Pieper et al., 2016). Excess protein in the diet increases undigested protein that contributes to undesirable fermentation in the large intestine by proteolytic bacteria, disrupting gut functionality in both young and adult pigs. Although results vary, several studies have investigated the effects of reducing dietary crude protein (CP) by 3% to 6% (compared to the NCR recommended level) in improving the health and intestinal eubiosis of pigs, when supplemented with amino acids. The positive effects of a low CP diet on intestinal health are related to promoting intestinal eubiosis and an increase of beneficial bacteria. Several studies have shown an increase in microbial diversity and a reduction of potentially pathogenic bacteria such as *Streptococcaceae* and *Enterobacteriaceae* in the small intestine, as well as a reduction in microbial diversity and an increase of *Lachnospiraceae*, *Prevotellaceae*, and *Veillonellaceae* in the large intestine of both growing and weaning pigs fed a reduced level of proteins (Chen et al., 2018b; Wan et al., 2020; Yu et al., 2019a). The shift of microbial composition modifies the production of VFAs and biogenic amine mainly in the large intestine. Apart from propionic acid and (to some extent) acetic acid, the amount of the other VFAs in the cecum is generally reduced in pigs fed a diet with a CP level lower than 17% (Fig. 1).

It has been suggested that reducing protein levels can improve intestinal mucosa absorption and development in young piglets. Data on ileum morphological parameters suggest that the effect depends on the level of protein reduction (Fig. 2). More promising results in improving the villus and reducing crypt depth can be obtained at 17% of CP, while 12–13% of CP can damage the intestinal mucosa, exerting a negative effect on the absorption of

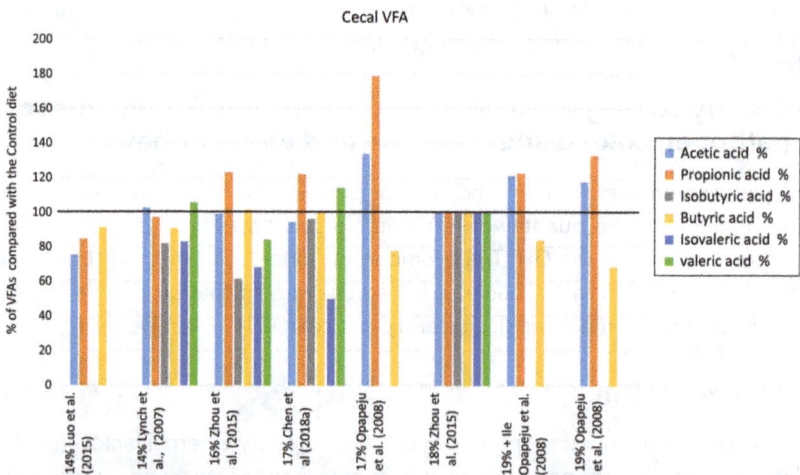

Figure 1 Percentage of acetic, propionic, Isobutyric, butyric, isovalenic and valenic acids in the cecum of pigs fed a diet with reduced amount of crude protein (CP) compared to the control diet. Level of fatty acids in the control diet (level pf CP ≥ 20) is shown as a black line (100%).

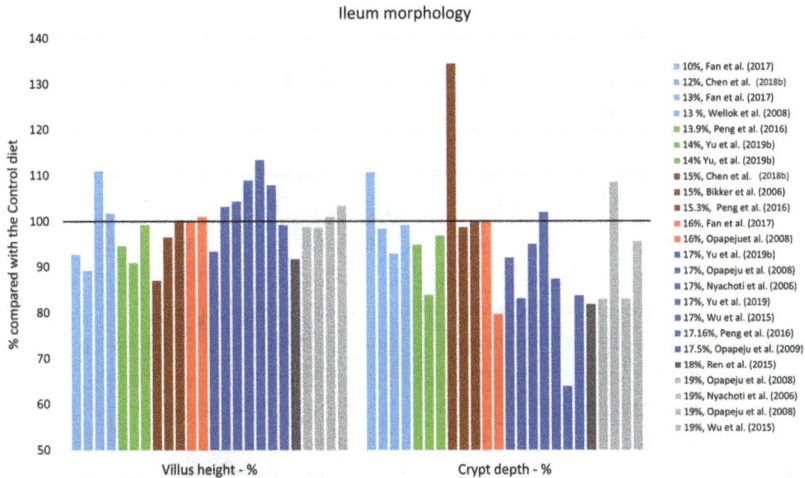

Figure 2 Percentage of villus height and crypt depth in the cecum of pigs fed a diet with reduced amount of crude protein (CP) compared to the control diet. Level of villus height and crypt depth in the control diet (level pf CP ≥ 20) is shown as a black line (100%).

nutrients. It has been observed that a consistent reduction of CP (10-12% CP) can significantly reduce the amount of biogenic amine, especially cadaverine and spermidine, that can protect the intestinal mucosa and the epithelial cells (Chen et al., 2018b; Fan et al., 2017).

Reducing the level of protein has been investigated as a potential strategy to reduce post-weaning diarrhea. A number of studies have observed a reduction of the fecal score and corresponding incidence of diarrhea in pigs fed a low protein diet (Wen et al., 2018; Opapeju et al., 2008; Lordelo et al., 2008; Nyachoti et al., 2006). *In vivo* challenge trials with ETEC F4 suggest a promising effect of lower protein diets in counteracting pathogen infection (Opapeju et al., 2015; Kim et al., 2011; Heo et al., 2009). However, some studies have suggested that a high protein diet does not always cause post-weaning diarrhea (PWD) (Htoo et al., 2007). Gilbert et al. (2019) have recently suggested that protein fermentation is associated with PWD but does not precede PWD, and that additional factors, besides protein fermentation metabolites, contribute to PWD development.

Grahofer et al. (2016) have studied subclinical infection of *Brachyspira hyodysenteriae* on 5 Swiss pig farms. Short-term substitution (2 days) of soybean meal (48% crude protein) in half the diet for growing pigs (less than 100 days of age) was followed by a decrease in dry matter content of feces sampled after 6 days, while dry matter increased in older fattening pigs. The number of positive *B. hyodysenteriae* samples was low (7 of 597). This suggests that the impact of dietary protein content on gut health can differ with age and, presumably, with different levels of adaptation of the gut or feed transit time.

6.2 Dietary fiber

Dietary fiber can significantly influence gut functionality. Dietary fiber includes heterogeneous class non-starch polysaccharides (NSPs) that provide a substrate for microbial fermentation. NSPs are normally divided into non-soluble (iNSP) fibers and soluble (sNSP) fibers, with sNSPs being more easily, rapidly and completely fermented. sNSPs contribute to modifying intestinal microbiota, reducing intestinal pathogens and thus the incidence of diarrhea, especially in weaning pigs. The contribution of iNSPs in reducing the incidence of diarrhea has been attributed to its high water-holding capacity (Montagne et al., 2003). Table 1 shows the most common feed ingredient sources of iNSPs and sNSPs used for weaning pigs (Molist et al., 2014).

The fermentability and viscosity of the digesta derived from dietary fiber can promote gut health by modulating the intestinal microenvironment. Fermentation of sNSPs starts in the ileum; in contrast, iNSPs increase the water-retention of intestinal contents and promote slower fermentation in the large intestine (Freire et al., 2000). The main results of DF fermentation are short-chain fatty acids (SCFAs) (acetate, propionate and butyrate), gases (CO_2, H_2 and CH_4) and other metabolites such as lactate that increase depending on the NSP level (Molist et al., 2009; Drochner et al., 2004). Studies in weaning pigs suggested that sNSPs are more frequently associated with PWD than iNSPs. Insoluble fiber promotes a beneficial shift of microbial colonization and prevents intestinal disease associated with *E. coli* and *B. pilosicoli* in weaned and growing pigs (Chen et al., 2014; Molist et al., 2009; Gerritsen et al., 2012; Thomsen et al., 2007) and of *S. suis* in suckling piglets (Zhang et al., 2016).

An sNPS diet increases the viscosity of digesta, reduces transit time along the intestine, affects intestinal contraction, and increases the thickness of the mucus layer (Knudsen, 2001). The increase of mucus may have beneficial effects: reducing the adhesion of bacteria to epithelial cells and promoting

Table 1 List of the most common feed ingredient sources of non-soluble fiber (iNSP) and soluble fiber (sNSP) in weaning pigs

iNSP	sNSP
alfalfa meal	inulin
barley hulls	sugar-beet pulp
oat hulls	pearl barley
cereal brans	citrus pulp
straw	soybean hulls
purified cellulose	
soybean hulls	

the development of opportunistic beneficial bacteria. The interaction between mucus (thickness and composition) and pathogens is still to be elucidated. It is known, for example, that *E. coli* can adhere to both epithelial cells and mucins. It has been suggested that the viscus of sNSPs promotes the proliferation of spirochete, including *B. pilosicoli* (Hopwood et al., 2002; Naresh and Hampson, 2010).

Besides the potential effect on bacteria, high viscosity digesta increase villus cell losses, leading to a reduction of villus length and an increase of crypt depth. Carboxymethylcellulose (CMC) increases viscosity in the small intestinal without producing SCFA. In contrast, the addition of low viscosity DF promotes small intestinal morphology (McDonald et al., 2001). Cereal polysaccharides such as arabinoxylans can also be considered as direct immunomodulators (Mendis et al., 2016).

One option is to combine fermentable proteins and fermentable carbohydrates since fermentable carbohydrates can reduce harmful protein fermentation in the porcine gut. Fermentable carbohydrates reduce the production of bacterial metabolites in the large intestine of pigs fed high fermentable proteins diets (Pieper et al., 2012; Stumpff et al., 2013). They also affect the expression of pro and anti-inflammatory genes and the expression of tight junctions (Pieper et al., 2012; Richter et al., 2014). However, they do not reduce the production of NH_3 and putrescine (Tudela et al., 2015). The inclusion of carbohydrates does not reduce intestinal inflammation and oxidative stress (Pieper et al., 2012). They do not increase the expression of monocarboxylate transporter 1 (*MCT1*) involved in intestinal epithelial cells promoting barrier function (Pieper et al., 2012; Richter et al., 2014). According to Richter et al., the adaptation of mucosa may occur in the intestinal environment.

7 Dietary strategies to improve gut functionality and prevent pathogen colonization: dietary nitrate, organic and amino acids

7.1 Dietary nitrate

Another strategy to promote intestinal endogenous secretion is related to dietary nitrate. Dietary nitrates are partially re-circulated from the gut to saliva and reduced to nitrite by oral bacteria. On returning to the stomach, they promote acid secretion, reducing *in vitro* survival of enteropathogen bacteria (Rao et al., 2006). However, the dietary addition of nitrate (2.45% KNO_3) did not reduce the excretion of *S. enterica* serovar Typhimurium, in weaning pigs (Bosi et al., 2007a), presumably due to limited recirculation of nitrate into saliva.

7.2 Organic acids

The gastric acidic function can be promoted by supplementing organic acids in the diet of young pigs (Partanen and Mroz, 1999). Gastric acid function at weaning is not sufficiently developed due to milk fermentation which produces lactic acid (Bolduan et al., 1988). Adding calcium formate reduced ETEC F4 infection in piglets (1.2% diet; Bosi et al., 2007b). However, it did not reduce total *E. coli* in the ileum in piglets challenged with ETEC K99 (1.8% diet, Torrallardona et al., 2007). A mixture of 75% formic acid and 25% propionic acid reduced gastric concentration of *Salmonella enterica* serovar Derby or hemolytic *E. coli* in weaned pigs (Taube et al., 2009).

Adding free calcium formate to the diet of weaning pigs reduced parietal cell population and gastric RNA expression for H+/K+-ATPase (Bosi et al., 2006). This slows activation of acid secretion, suggesting the need for a gradual shift to unsupplemented diets as pigs grow. The use of fat-protected calcium formate also has a protective effect against ETEC F4 without impairing the development of gastric fundic mucosa (Bosi et al., 2007b).

Another option is the use of potassium diformate which reduces the number of coliforms and *Streptococci* in the stomach and proximal colon, and the number of coliforms in the colon (Février et al., 2001), shedding the *Salmonella enterica* serovar Derby (Taube et al., 2009). Potassium diformate did not affect maturation of gastric acidic function; it did not change the concentration of gastrin and the activity of H+-K+-ATPase in the gastric oxyntic mucosa, whilst upregulating mRNA expression of gastrin and H+-K+-ATPase, notwithstanding decreased gastric hydrochloric acid concentration (Xia et al., 2016).

7.3 Amino acids

Adding amino acids may regulate endogenous defenses against enteropathogens. A review of the potential role of functional amino acids in supporting the health of growing pigs can be found in Le Floc'h et al. (2018). A total of 0.4% isoleucine supplementation (70 % isoleucine to lysine ratio) of the standard diet to piglets weaned at 21 days upregulated mRNA expression of several porcine β-defensins in the jejunum and ileum (Ren et al., 2019). This was associated with a reduction of plasmatic values of lipopolysaccharide endotoxin by isoleucine within ETEC-challenged pigs, showing that the amino acid reduced the effect of the infection.

Threonine is an amino acid contained in glycoproteins in porcine small-intestinal mucus (Mantle and Allen, 1981). It also is an essential amino acid in immunoglobulins (Le Floc'h et al., 2018). Increasing the dietary ratio of standard ileal digestible (SID) threonine to SID lysine from 65% to 70% increased the IgA ETEC-specific activity in plasma and jejunal secretion of ETEC-susceptible subjects but not of ETEC-resistant subjects (Trevisi et al., 2015b). However, no improvement in the total mucin content in jejunal mucosal scrapings and in mucin-secreting goblet cell numbers was seen (Trevisi et al., 2015b). Increasing SID threonine to lysine ratios up to 71% increased the ileal mRNA for mucin 20 but not for mucin 1 and mucin 4 (Jayaraman, 2019). Dietary threonine may prioritize the production of mucins among proteins with high threonine content. This is suggested by the maintenance of protein synthesis in mucin-producing gastro-intestinal tracts at the expense of muscle and other tissues in mini-pigs when the diet was 20-220% of threonine requirements (Munasinghe et al., 2017). Nevertheless, given that *Salmonella* infection stimulates mucin production, the threonine requirements of piglets should be carefully assessed (Wellington et al., 2020).

Glutamine can be important in situations where local tissue activation is required (e.g., when the intestine is challenged). It is an energy source and a precursor for the synthesis of purine and pyrimidine nucleotides (Le Floc'h et al., 2018). Supplementing the diet of early-weaned piglets with glutamine preserves normal lymphocyte function following *E. coli* infection (Yoo et al., 1997). It also attenuates mucosal cytokine response (Ewaschuk et al., 2011) as well as ameliorating fecal score and villus morphology (Yi et al., 2005). Glutamine can thus reduce the severity of ETEC infection.

Amino acids are also important nutrients for several bacteria and can create selective opportunities for enteric pathogens, although there is no direct evidence of the effect of a particular amino acid in pigs (Dai et al., 2010). Increasing L-serine in mice promoted invasive *E. coli* LF82 (Kitamoto et al., 2020). Lowering serine reduced the competitive advantage of the pathogen and intestinal inflammation. However, serine is important for O-glycosylation of the mucins. The addition of 0.2% of L-serine in weaned pigs upregulated marker genes of tight junction function in the small intestine, reduced inflammatory response and improved villus development (Zhou et al., 2018). There is a need for more research on the association of dietary amino acids (supplemented alone or in dietary proteins) with commensal and pathogen colonization.

8 Dietary interventions for pathogen-specific defence

Intestinal pathogens use different molecules on the host intestinal surface to forage, to avoid being flushed away and to exploit their pathogenicity (see the earlier discussion of host genotype effect). The mechanism of adhesion is quite complex, for example, the recognition mechanism between host glycosylation motifs and bacteria appendages. A general strategy is to block the forming of the bond between the pathogen and the host surface. However, due to the specificity of the bond, very accurate tests are required to select the molecules or products that can limit opportunities for colonization by the microbe.

The availability of the brush border of porcine intestinal villi of F4- or F18-ETEC susceptible pigs provides the opportunity to test soluble products. Trevisi et al. (2012) screened yeast cell wall-based products capable of avoiding ETEC F4ac adhesion by sequestering the microbe. In addition, products capable of competing on the host surface can mimic bacterial fimbriae that have properties like lectins. Galactose motifs may be involved in the adhesion of F4 ETEC and, among a selection of d-galactose lectins of vegetal origin, a few were able to avoid *in vitro* pathogen-specific adhesion to pig brush border receptors (Trevisi et al., 2017b).

A range of polyphenol extracts have been tested *in vitro* against F4 ETEC (Verhelst et al. (2010). These included: extract from cocoa beans (Omnicoa 45), pentagalloyl glucose crude (PGG) and purified hydrolysable tannin consisting of medium to high molecular weight polyphenols (ALSOK). ALSOK and PGG inhibited LT toxins at a minimum concentration of 750 µg/mL. All products demonstrated inhibitory action against ETEC in brush border adhesion tests. In an *in vivo* trial, Verhelst et al. (2014) supplemented the diet of weaned pigs challenged with F4 ETEC with 1% of ALSOK, omnicoa or extract from grape seeds (condensed tannins – Omnivin). Whilst ALSOK had no effect, Omnivin reduced ETEC excretion in pigs but reduced the average daily gain.

There have been studies of ways of blocking ETEC heat-stable toxin (ST), associated with chloride-rich liquid diarrheic secretion induced by cGMP cascade. When bound to zinc, ETEC was not able to induce cGMP *in vitro* (Kiefer et al., 2020). These observations suggest zinc can protect Caco-2 intestinal culture cells against ETEC adhesion and internalization, preventing the increase of tight junction permeability (Roselli et al., 2003). The antimicrobial effect of zinc oxide explains the use of zinc supplementation in piglet feeding. This chapter does not discuss in detail the growth-promoting effect of zinc oxide supplementation in piglet feed (3000 mg/kg feed, Yu et al., 2017 for more information). Some countries (e.g. those of the EU) have banned the practice given potential pollution from zinc in the effluent. Some zinc formulations (coated zinc oxide, Kim et al., 2015; Lei and Kim, 2020; or zinc glutamate chelated, Bosi et al., 2003) have shown promising results against enteropathogens. Lower doses of zinc can reduce ETEC shedding, the severity of diarrhea, inflammation response and deterioration of intestinal morphology induced by ETEC K88 in weaned pigs. More studies are needed to balance the beneficial zinc effect on the gut and zinc losses to the environment.

Bacteria secrete molecules to achieve a competitive growth advantage. Colicinogenic *E. coli* reduced *E. coli* counts from ileum mucosa of early-weaned pigs challenged with ETEC (Bhandari et al., 2010). *Lactobacillus plantarum* ZLP001 has been shown to inhibit ETEC growth (Wang et al., 2018).

Beneficial microbes may use the same molecular motifs as enteropathogens to adhere to mucous surfaces or intestinal epithelium. *In vitro* adherence of *Salmonella, E. coli* and *Clostridum* spp. to the porcine intestinal mucosa were reduced by *Bifidobacter lactis* or *Lactobacillus rhamnosus* (Collado et al., 2007). The adhering ability of *L. plantarum* ZLP001 IPEC1 prevented ETEC adhesion (Wang et al., 2018). This was associated with the activation of the innate immune response to antimicrobial peptides, through recognition by the intestinal bacterial receptor TLR2 (Wang et al., 2019).

Beneficial bacteria are able to either activate specific host defense mechanisms or down-regulate evasion mechanisms induced by pathogens. Dietary provision of *Bacillus cereus* var. Toyoi in piglets challenged with *S. enterica* serovar Typhimurium DT104 reduced the number of CD8+ γδ T cells in peripheral blood and the jejunal epithelium and reduced diarrhea (Scharek-Tedin et al., 2013). Among other defense molecules produced by the host, lysozyme is found in digestive secretions and in colostrum and milk. Supplementing piglets from 3 days of age with milk from transgenic sows expressing recombinant human lysozyme reduced the severity and duration of diarrhea after being challenged with ETEC F4 (Huang et al., 2018). Supplementing weaning pigs with probiotics stimulates the growth of local lymphocytes which may be associated with reduced signs of illness. Feeding probiotics for 1 week after weaning upregulated the gene sets involved in the differentiation and proliferation of T and B lymphocytes (Trevisi et al., 2017a), and this contributed to reducing diarrhea in the first few days after infection and delayed mortality (Trevisi et al., 2015a).

Feeding pre- and post-weaning piglets *Enterococcus faecium* NCIMB 10415 (including indirectly via supplementing feed to their mothers) stimulated intraepithelial lymphocytes and villous morphometry after 4 weeks of infection with *S. enterica* serovar Typhimurium (Rieger et al., 2015). However, there was more concentration of *Salmonella* in tonsils and no effect on fecal *Salmonella* shedding, due to an anti-inflammatory/immuno-suppressive effect of *Enterococcus faecium* being more evident in post-weaning piglets (Siepert et al., 2014). Bacteria can also secrete molecules that disrupt cell membranes, as seen in the effect of *Bacillus pumilus* H2 on Vibrio (Gao et al., 2017). When added to a weaning pig diet, this explains the ability of a *B. pumilus* strain to reduce *L. intracellularis* shedding and lesions, while *B. amyloliquefaciens* was ineffective (Opriessnig et al., 2019). More studies are needed to better understand the specific action of beneficial bacteria on pigs exposed to pathogens. A positive effect of a probiotic on pigs' growth does not necessarily imply protection against enteric pathogens. *Bifidobacterium animalis* RA18 had promoted growth in healthy weaning pigs (Modesto et al., 2009) but was not able to combat the effects of F4 ETEC (Modesto et al., 2007).

9 Future trends and conclusion

Disentangling the physiological basis of resilience to disease in pigs is key to more sustainable production. This chapter highlights the key gaps in research to improve gut function in pigs in preventing gut pathogen colonization. Considering the 3Rs (replacement, refinement and reduction) principles in animal research, there is the need to develop *in vitro* and *ex vivo* tests able to replicate *in vivo* conditions to improve the translational potential of results obtained using animal models.

The pre-weaning period is well recognized as an important phase in which well-targeted interventions able to influence the first stage of adaptation to the extra-uterine environment can be applied. This has the potential to modulate gut maturation programming and then the physiological development of piglets, impacting the natural resistance of pigs against enteropathogen microbes. The post-weaning period deserves further attention, especially focusing on in-water supplementation, to reduce the physiological stress due to reduced feed intake in the first days of the post-weaning period. Finally, there is a need to develop next-generation sequencing (NGS) tools to better trace gut traits in the riskiest phases of pig production. This will support the design of new dietary strategies to increase the robustness of pigs against colonization and multiplication of enteropathogenic microbes.

10 Where to look for further information

Seminal articles on the topic:

- Huting, A., Middelkoop, A., Guan, X. and Molist, F. (2021). Using nutritional strategies to shape the gastro-intestinal tracts of suckling and weaned piglets. *Animals*, 11(2), 402.
- Lalles, J. P., Bosi, P., Smidt, H. and Stokes, C. R. (2007). Nutritional management of gut health in pigs around weaning. *Proceedings of the Nutrition Society*, 66(2), 260–268.
- Pieper, R., Tudela, C. V., Taciak, M., Bindelle, J., Pérez, J. F. and Zentek, J. (2016). Health relevance of intestinal protein fermentation in young pigs. *Animal Health Research Reviews*, 17(2), 137–147.
- Pluske, J. R., Pethick, D. W., Hopwood, D. E. and Hampson, D. J. (2002). Nutritional influences on some major enteric bacterial diseases of pig. *Nutrition Research Reviews*, 15(2), 333–371.
- Trevisi, P., Luise, D., Correa, F. and Bosi, P. (2021). Timely control of gastrointestinal eubiosis: a strategic pillar of pig health. *Microorganisms*, 9(2), 313.

11 References

Albera, E. and Kankofer, M. (2009). Antioxidants in colostrum and milk of sows and cows, *Reprod. Domest. Anim.* 44(4), 606-611.

Alvarez-Ordóñez, A., Martinez-Lobo, F. J., Arguello, H., Carvajal, A. and Rubio, P. (2013). Swine dysentery: aetiology, pathogenicity, determinants of transmission and the fight against the disease, *Int. J. Env. Res. Pub. Health* 10, 1927-1947.

Andres, V. M. and Davies, R. H. (2015). Biosecurity measures to control Salmonella and other infectious agents in pig farms: a review, *Compr. Rev. Food Sci. F.* 14(4), 317-335.

Aumaitre, A., Peiniau, J. and Madec, F. (1995). Digestive adaptation after weaning and nutritional consequences in the piglet, *Pig News Inform.* 16, 73N-79N.

Aumiller, T., Mosenthin, R. and Weiss, E. (2015). 'Potential of cereal grains and grain legumes in modulating pigs' intestinal microbiota - a review, *Livest. Sci.* 172, 16-32.

Bäckhed, F. (2011). Programming of host metabolism by the gut microbiota, *Ann. Nutr. Metab.* 58 (Suppl. 2), 44-52.

Bhandari, S. K., Opapeju, F. O., Krause, D. O. and Nyachoti, C. M. (2010). Dietary protein level and probiotic supplementation effects on piglet response to Escherichia coli K88 challenge: performance and gut microbial population, *Livest. Sci.* 133(1-3), 185-188.

Bikker, P., Dirkzwager, A., Fledderus, J., Trevisi, P., Le Huërou-Luron, I., Lallès, J. P. and Awati, A. (2006). The effect of dietary protein and fermentable carbohydrates levels on growth performance and intestinal characteristics in newly weaned piglets, *J. Anim. Sci.* 84(12), 3337-3345.

Bolduan, V. G., Jung, H., Schnabel, E. and Schneider, R. (1988). Recent advances in the nutrition of weaner piglets, *Pig News Inf.* 9(4), 381-385.

Bosi, P., Merialdi, G., Sarli, G., Casini, L., Gremokolini, C., Preziosi, R., Brunetti, B. and Trevisi, P. (2003). Effects of doses of ZnO or Zn-glutamate on growth performance, gut characteristics, health and immunity of early-weaned pigs orally challenged with E. coli K88, *Ital. J. Anim. Sci.* 2(1), 361-363.

Bosi, P., Mazzoni, M., De Filippi, S., Trevisi, P., Casini, L., Petrosino, G. and Lalatta-Costerbosa, G. (2006). A continuous dietary supply of free calcium formate negatively affects the parietal cell population and gastric RNA expression for H+/K+-ATPase in weaning pigs, *J. Nutr.* 136(5), 1229-1235.

Bosi, P., Casini, L., Tittarelli, C., Minieri, L., De Filippi, S., Trevisi, P., Clavenzani, P. and Mazzoni, M. (2007a). Effect of dietary addition of nitrate on growth, salivary and gastric function, immune response, and excretion of Salmonella enterica serovar Typhimurium, in weaning pigs challenged with this microbe strain, *Ital. J. Anim. Sci.* 6(1), 266-268.

Bosi, P., Sarli, G., Casini, L., De Filippi, S., Trevisi, P., Mazzoni, M. and Merialdi, G. (2007b). The influence of fat protection of calcium formate on growth and intestinal defence in Escherichia coli K88-challenged weanling pigs, *Anim. Feed Sci. Technol.* 139(3-4), 170-185.

Boye, M., Baloda, S. B., Leser, T. D. and Møller, K. (2001). Survival of Brachyspira hyodysenteriae and B. pilosicoli in terrestrial microcosms, *Vet. Microbiol.* 81(1), 33-40.

Burrough, E. R. (2017). Swine dysentery: etiopathogenesis and diagnosis of a re-emerging disease, *Vet. Pathol.* 54(1), 22-31.

Carstensen, L., Ersbøll, A. K., Jensen, K. H. and Nielsen, J. P. (2005). Escherichia coli post-weaning diarrhoea occurrence in piglets with monitored exposure to creep feed, *Vet. Microbiol.* 110(1-2), 113-123.

Chen, H., Mao, X. B., Che, L. Q., Yu, B., He, J., Yu, J., Han, G. Q., Huang, Z. Q., Zheng, P. and Chen, D. W. (2014). Impact of fiber types on gut microbiota, gut environment and gut function in fattening pigs, *Anim. Feed Sci. Technol.* 195, 101-111.

Chen, W., Mi, J., Lv, N., Gao, J., Cheng, J., Wu, R., Ma, J., Lan, T. and Liao, X. (2018a). Lactation stage-dependency of the sow milk microbiota, *Front. Microbiol.* 9, 945.

Chen, X., Song, P., Fan, P., He, T., Jacobs, D., Levesque, C. L., Johnston, L. J., Ji, L., Ma, N., Chen, Y., Zhang, J., Zhao, J. and Ma, X. (2018b). Moderate dietary protein restriction optimized gut microbiota and mucosal barrier in growing pig model, *Front. Cell. Infect. Microbiol.* 8, 246.

Chen, X., Xu, J., Ren, E., Su, Y. and Zhu, W. (2018c). Co-occurrence of early gut colonization in neonatal piglets with microbiota in the maternal and surrounding delivery environments, *Anaerobe* 49, 30-40.

Cheng, L., Xu, Q., Yang, K., He, J., Chen, D., Du, Y. and Yin, H. (2016). Annotation of porcine milk oligosaccharides throughout lactation by hydrophilic interaction chromatography coupled with quadruple time of flight tandem mass spectrometry, *Electrophoresis* 37(11), 1525-1531.

Chouet, S., Pietro, C., Mieli, L., Veenhuizen, M. F. and McOrist, S. (2003). Patterns of exposure to Lawsonia intracellularis infection on European pig farms, *Vet. Rec.* 152(1), 14-17.

Coddens, A., Verdonck, F., Mulinge, M., Goyvaerts, E., Miry, C., Goddeeris, B., Duchateau, L. and Cox, E. (2008). The possibility of positive selection for both F18+ Escherichia coli and stress resistant pigs opens new perspectives for pig breeding, *Vet. Microbiol.* 126(1-3), 210-215.

Collado, M. C., Grześkowiak, Ł. and Salminen, S. (2007). Probiotic strains and their combination inhibit in vitro adhesion of pathogens to pig intestinal mucosa, *Curr. Microbiol.* 55(3), 260-265.

Dai, Z. L., Zhang, J., Wu, G. and Zhu, W. Y. (2010). Utilization of amino acids by bacteria from the pig small intestine, *Amino Acids* 39(5), 1201-1215.

de Oliveira, A. C. D. F., Vanelli, K., Sotomaior, C. S., Weber, S. H. and Costa, L. B. (2019). Impacts on performance of growing-finishing pigs under heat stress conditions: a meta-analysis, *Vet. Res. Commun.* 43(1), 37-43.

Dors, A., Pomorska-Mól, M., Czyżewska, E., Wasyl, D. and Pejsak, Z. (2015). Prevalence and risk factors for *Lawsonia intracellularis*, *Brachyspira hyodysenteriae* and *Salmonella* spp. in finishing pigs in Polish farrow-to-finish swine herds, *Pol. J. Vet. Sci.* 18, 825-831.

Drochner, W., Kerler, A. and Zacharias, B. (2004). Pectin in pig nutrition, a comparative review, *J. Anim. Physiol. Anim. Nutr. (Berl)* 88(11-12), 367-380.

Dubreuil, J. D., Isaacson, R. E. and Schifferli, D. M. (2016). Animal enterotoxigenic *Escherichia coli, EcoSal Plus* 7(1), doi: 10.1128/ecosalplus.ESP-0006-2016.

Duhamel, G. E. (1998). Colonic spirochetosis caused by *Serpulina pilosicoli, Large Anim. Pract.* 19, 14-22.

Duhamel, G. E. (2001). Comparative pathology and pathogenesis of naturally acquired and experimentally induced colonic spirochetosis, *Anim. Health Res. Rev.* 2(1), 3-17.

Edfors-Lilja, I., Wattrang, E., Magnusson, U. and Fossum, C. (1994). Genetic variation in parameters reflecting immune competence of swine, *Vet. Immunol. Immunopathol.* 40(1), 1-16.

EFSA (European Food Safety Authority) (2009). Analysis of the baseline survey on the prevalence of Salmonella in holdings with breeding pigs in the EU, 2008 - part A: Salmonella prevalence estimates, *EFSA J.* 7(12), 1377.

EFSA (European Food Safety Authority) and ECDC (European Centre for Disease Prevention and Control) (2016). The European Union summary report on trends and sources of zoonoses, zoonotic agents and food-borne outbreaks in 2015, *EFSA J.* 13, 4329.

Elliot, J. I., Senft, B., Erhardt, G. and Fraser, D. (1984). 'Isolation of lactoferrin and its concentration in sows' colostrum and milk during a 21-day lactation, *J. Anim. Sci.* 59(4), 1080-1084.

Ewaschuk, J. B., Murdoch, G. K., Johnson, I. R., Madsen, K. L. and Field, C. J. (2011). Glutamine supplementation improves intestinal barrier function in a weaned piglet model of Escherichia coli infection, *Br. J. Nutr.* 106(6), 870-877.

Fan, P., Liu, P., Song, P., Chen, X. and Ma, X. (2017). Moderate dietary protein restriction alters the composition of gut microbiota and improves ileal barrier function in adult pig model, *Sci. Rep.* 7, 43412.

Fedorka-Cray, P. J. and Wray, C. (2000). Salmonella infections in pigs. In: Wray, C. and Wray, A. (Eds), *Salmonella in Domestic Animals*, London: CAB International, pp. 191-207.

Février, C., Gotterbarm, G., Jaghelin-Peyraud, Y., Lebreton, Y., Legouevec, F. and Aumaitre, A. (2001). Effects of adding potassium diformate and phytase excess for weaned piglet. In: Lindberg, J. E. and Ogle, B. (Eds), *Digestive Physiology of Pigs*, London: CABI publishing, pp. 136-138.

Freire, J. P. B., Guerreiro, A. J. G., Cunha, L. F. and Aumaitre, A. (2000). Effect of dietary fibre source on total tract digestibility, caecum volatile fatty acids and digestive transit time in the weaned piglet, *Anim. Feed Sci. Technol.* 87(1-2), 71-83.

Gabler, N. K., Koltes, D., Schaumberger, S., Murugesan, G. R. and Reisinger, N. (2018). Diurnal heat stress reduces pig intestinal integrity and increases endotoxin translocation, *Transl. Anim. Sci.* 2(1), 1-10.

Galina-Pantoja, L., Siggens, K., Van Schriek, M. G. M. and Heuven, H. C. M. (2009). Mapping markers linked to porcine salmonellosis susceptibility, *Anim. Genet.* 40(6), 795-803.

Gao, X. Y., Liu, Y., Miao, L. L., Li, E. W., Hou, T. T. and Liu, Z. P. (2017). Mechanism of anti-Vibrio activity of marine probiotic strain Bacillus pumilus H2, and characterization of the active substance, *AMB Express* 7(1), 23.

Geraci, C., Varzandi, A. R., Schiavo, G., Bovo, S., Ribani, A., Utzeri, V. J., Galimberti, G., Buttazzoni, L., Ovilo, C., Gallo, M., Dall'Olio, S. and Fontanesi, L. (2019). Genetic markers associated with resistance to infectious diseases have no effects on production traits and haematological parameters in Italian Large White pigs, *Livest. Sci.* 223, 32-38.

Gerritsen, R., Van Der Aar, P. and Molist, F. (2012). Insoluble nonstarch polysaccharides in diets for weaned piglets, *J. Anim. Sci.* 90(4), 318-320.

Gilbert, M. S., van der Hee, B. and Gerrits, W. J. J. (2019). The role of protein fermentation metabolites in post-weaning diarrhoea in piglets. In: *EAAP Scientific Series*, Wageningen: Wageningen Academic Publishers, pp. 661-667. (Book of abstracts No. 25, Ghent, Belgium, 26-30 August 2019).

Goetstouwers, T., Van Poucke, M., Coppieters, W., Nguyen, V. U., Melkebeek, V., Coddens, A., Van Steendam, K., Deforce, D., Cox, E. and Peelman, L. J. (2014). Refined candidate region for F4ab/ac enterotoxigenic Escherichia coli susceptibility situated proximal to MUC13 in pigs, *PLoS ONE* 9(8), e105013.

Grahofer, A., Overesch, G., Nathues, H. and Zeeh, F. (2016). Effect of soy on faecal dry matter content and excretion of Brachyspira hyodysenteriae in pigs, *Vet. Rec. Open* 3(1), e000159.

Gray, J. T., Fedorka-Cray, P. J., Stabel, T. J. and Ackermann, M. R. (1995). Influence of inoculation route on the carrier state of Salmonella choleraesuis in swine, *Vet. Microbiol.* 47(1-2), 43-59.

Hampson, D. J., La, T. and Phillips, N. D. (2015). Emergence of Brachyspira species and strains: reinforcing the need for surveillance, *Porc. Health Manag.* 1(1), 8.

Hampson, D. J. and Burrough, E. R. (2019). Swine dysentery and Brachyspiral colitis. In: Zimmerman, J. J., Karriker, L. A., Ramirez, A., Schwartz, K. J., Stevenson, G. W. and Zhang, J. (Eds), *Diseases of Swine*, Hoboken: Wiley Online Editions, pp. 951-970.

Harris, N. L., Spoerri, I., Schopfer, J. F., Nembrini, C., Merky, P., Massacand, J., Urban, J. F., Lamarre, A., Burki, K., Odermatt, B., Zinkernagel, R. M. and Macpherson, A. J. (2006). Mechanisms of neonatal mucosal antibody protection, *J. Immunol.* 177(9), 6256-6262.

Hedemann, M. S., Mikkelsen, L. L., Naughton, P. J. and Jensen, B. B. (2005). Effect of feed particle size and feed processing on morphological characteristics in the small and large intestine of pigs and on adhesion of Salmonella enterica serovar Typhimurium DT12 in the ileum in vitro, *J. Anim. Sci.* 83(7), 1554-1562.

Heo, J. M., Kim, J. C., Hansen, C. F., Mullan, B. P., Hampson, D. J. and Pluske, J. R. (2009). Feeding a diet with decreased protein content reduces indices of protein fermentation and the incidence of postweaning diarrhea in weaned pigs challenged with an enterotoxigenic strain of Escherichia coli, *J. Anim. Sci.* 87(9), 2833-2843.

Hodgins, D. C., Kang, S. Y., De Arriba, L., Parreño, V., Ward, L. A., Yuan, L., To, T. and Saif, L. J. (1999). Effects of maternal antibodies on protection and development of antibody responses to human rotavirus in gnotobiotic pigs, *J. Virol.* 73(1), 186-197.

Hopwood, D. E., Pethick, D. W. and Hampson, D. J. (2002). Increasing the viscosity of the intestinal contents stimulates proliferation of enterotoxigenic *Escherichia coli* and *Brachyspira pilosicoli* in weaner pigs, *Br. J. Nutr.* 88(5), 523-532.

Htoo, J. K., Araiza, B. A., Sauer, W. C., Rademacher, M., Zhang, Y., Cervantes, M. and Zijlstra, R. T. (2007). Effect of dietary protein content on ileal amino acid digestibility, growth performance, and formation of microbial metabolites in ileal and cecal digesta of early-weaned pigs, *J. Anim. Sci.* 85(12), 3303-3312.

Huang, G., Li, X., Lu, D., Liu, S., Suo, X., Li, Q. and Li, N. (2018). Lysozyme improves gut performance and protects against enterotoxigenic Escherichia coli infection in neonatal piglets, *Vet. Res.* 49(1), 20.

Hurley, W. L. (2015). Composition of sow colostrum and milk. In: Farmer, C. (Ed), *The Gestating and Lactating Sow*, Wageningen: Wageningen Academic Publishers, pp. 115-127.

Jacobson, M., Andersson, M., Lindberg, R., Fossum, C. and Jensen-Waern, M. (2011). Microarray and cytokine analyses of field cases of pigs with diarrhoea, *Vet. Microbiol.* 153(3-4), 307-314.

Jayaraman, B. (2019). Evaluation of standardized ileal digestible threonine to lysine ratio and tryptophan to lysine ratio in weaned pigs fed antibiotic-free diets and subjected to immune challenge, Graduate Thesis, University of Manitoba.

Jha, R. and Berrocoso, J. F. D. (2016). Dietary fiber and protein fermentation in the intestine of swine and their interactive effects on gut health and on the environment: a review, *Anim. Feed Sci. Technol.* 212, 18-26.

Ji, H., Ren, J., Yan, X., Huang, X., Zhang, B., Zhang, Z. and Huang, L. (2011). The porcine MUC20 gene: molecular characterization and its association with susceptibility to enterotoxigenic Escherichia coli F4ab/ac, *Mol. Biol. Rep.* 38(3), 1593-1601.

Joling, P., Wever, P. J. M., Oskam, J. P. H. and Henken, A. M. (1993). Lymphocyte stimulation by phytohaemagglutinin and conca- navalin A in different swine breeds, *Livest. Prod. Sci.* 53, 341–350.

Jørgensen, C. B., Cirera, S., Anderson, S. I., Archibald, A. L., Raudsepp, T., Chowdhary, B., Edfors-Lilja, I., Andersson, L. and Fredholm, M. (2003). Linkage and comparative mapping of the locus controlling susceptibility towards *E. coli* F4ab/ac diarrhoea in pigs, *Cytogenet. Genome Res.* 102(1–4), 157–162.

Kadhim, H. M. (2020). Review of pathogenicity and virulence determinants in Salmonella, *EurAsian J. Biosci.* 14(1), 377–381.

Kelly, D., O'Brien, J. J. and McCracken, K. J. (1990a). Effect of creep feeding on the incidence, duration and severity of post-weaning diarrhoea in pigs, *Res. Vet. Sci.* 49(2), 223–228.

Kelly, D., Smyth, J. A. and McCracken, K. J. (1990b). Effect of creep feeding on structural and functional changes of the gut of early weaned pigs, *Res. Vet. Sci.* 48(3), 350–356.

Kich, J. D., Uthe, J. J., Benavides, M. V., Cantão, M. E., Zanella, R., Tuggle, C. K. and Bearson, S. M. (2014). TLR4 single nucleotide polymorphisms (SNPs) associated with Salmonella shedding in pigs', *J. Appl. Genet.* 55(2), 267–271.

Kiefer, M. C., Motyka, N. I., Clements, J. D. and Bitoun, J. P. (2020). Enterotoxigenic *Escherichia coli* heat-stable toxin increases the rate of zinc release from metallothionein and is a zinc-and iron-binding peptide, *mSphere* 5(2), e00146-20.

Kim, J. C., Heo, J. M., Mullan, B. P. and Pluske, J. R. (2011). Efficacy of a reduced protein diet on clinical expression of post-weaning diarrhoea and life-time performance after experimental challenge with an enterotoxigenic strain of Escherichia coli, *Anim. Feed Sci. Technol.* 170(3–4), 222–230.

Kim, S. J., Kwon, C. H., Park, B. C., Lee, C. Y. and Han, J. H. (2015). Effects of a lipid-encapsulated zinc oxide dietary supplement, on growth parameters and intestinal morphology in weanling pigs artificially infected with enterotoxigenic Escherichia coli, *J. Anim. Sci. Technol.* 57, 4.

Kitamoto, S., Alteri, C. J., Rodrigues, M., Nagao-Kitamoto, H., Sugihara, K., Himpsl, S. D., Bazzi, M., Miyoshi, M., Nishioka, T., Hayashi, A., Morhardt, T. L., Kuffa, P., Grasberger, H., El-Zaatari, M., Bishu, S., Ishii, C., Hirayama, A., Eaton, K. A., Dogan, B., Simpson, K. W., Inohara, N., Mobley, H. L. T., Kao, J. Y., Fukuda, S., Barnich, N. and Kamada, N. (2020). Dietary l-serine confers a competitive fitness advantage to Enterobacteriaceae in the inflamed gut, *Nat. Microbiol.* 5(1), 116–125.

Knudsen, K. B. (2001). The nutritional significance of "dietary fibre" analysis, *J. Anim. Sci. Technol.* 90(1–2), 3–20.

Krakowski, L., Krzyżanowski, J., Wrona, Z., Kostro, K. and Siwicki, A. K. (2002). The influence of nonspecific immunostimulation of pregnant sows on the immunological value of colostrum, *Vet. Immunol. Immunopathol.* 87(1–2), 89–95.

Kroll, J. J., Roof, M. B., Hoffman, L. J., Dikson, J. S. and Harris, D. L. H. (2005). Proliferative enteropathy: a global enteric disease of pigs caused by Lawsonia intracellularis, *Anim. Health Res. Rev.* 6(2), 173–197.

Kubasova, T., Davidova-Gerzova, L., Merlot, E., Medvecky, M., Polansky, O., Gardan-Salmon, D., Quesnel, L. and Rychlik, I. (2017). Housing systems influence gut microbiota composition of sows but not of their piglets, *PLoS ONE* 12(1), e0170051.

Lallès, J. P. and David, J. C. (2011). Fasting and refeeding modulate the expression of stress proteins along the gastrointestinal tract of weaned pigs, *J. Anim. Physiol. Anim. Nutr. (Berl)* 95(4), 478–488.

Lawson, G. H. K. and Gebhart, C. J. (2000). Proliferative enteropathy, *J. Comp. Pathol.* 122(2-3), 77-100.

Le Dividich, J. L., Rooke, J. A. and Herpin, P. (2005). Nutritional and immunological importance of colostrum for the new-born pig, *J. Agric. Sci.* 143(6), 469-485.

Le Floc'h, N., Wessels, A., Corrent, E., Wu, G. and Bosi, P. (2018). The relevance of functional amino acids to support the health of growing pigs, *J. Anim. Sci. Technol.* 245, 104-116.

Lee, S. H., Hosseindoust, A. R., Kim, J. S., Choi, Y. H., Lee, J. H., Kwon, I. K. and Chae, B. J. (2016). Bacteriophages as a promising anti-pathogenic option in creep-feed for suckling piglets: targeted to control Clostridium spp. and coliforms faecal shedding, *Livest. Sci.* 191(1), 161-164.

Lei, X. J. and Kim, I. H. (2020). Evaluation of coated zinc oxide in young pigs challenged with enterotoxigenic *Escherichia coli* K88, *J. Anim. Sci. Technol.* 262, 114399.

Levast, B., Berri, M., Wilson, H. L., Meurens, F. and Salmon, H. (2014). Development of gut immunoglobulin A production in piglet in response to innate and environmental factors, *Dev. Comp. Immunol.* 44(1), 235-244.

Liang, G., Zhao, C., Zhang, H., Mattei, L., Sherrill-Mix, S., Bittinger, K., Kessler, L. R., Wu, G. D., Baldassano, R. N., Derusso, P., Ford, E., Elovitz, M. A., Kelly, M. S., Patel, M. Z., Mazhani, T., Gerber, J. S., Kelly, A., Zemel. B. S. and Bushman, F. D. (2020). The stepwise assembly of the neonatal virome is modulated by breastfeeding. *Nature*, 581(7809), 470-474.

Liesner, V. G., Taube, V., Leonhard-Marek, S., Beineke, A.. and Kamphues, J. (2009). Integrity of gastric mucosa in reared piglets - effects of physical form of diets (meal/pellets), pre-processing grinding (coarse/fine) and addition of lignocellulose (0/2.5 %), *J. Anim. Physiol. Anim. Nutr.* 93(3), 373-380.

Liu, F., Cottrell, J. J., Furness, J. B., Rivera, L. R., Kelly, F. W., Wijesiriwardana, U., Pustovit, R. V., Fothergill, L. J., Bravo, D. M., Celi, P., Leury, B. J., Gabler, N. K. and Dunshea, F. R. (2016). Selenium and vitamin E together improve intestinal epithelial barrier function and alleviate oxidative stress in heat-stressed pigs, *Exp. Physiol.* 101(7), 801-810.

Liu, L., Wang, J., Zhao, Q., Zi, C., Wu, Z., Su, X., Huo, Y., Zhu, G., Wu, S. and Bao, W. (2013). Genetic variation in exon 10 of the BPI gene is associated with Escherichia coli F18 susceptibility in Sutai piglets, *Gene* 523(1), 70-75.

Liu, X. D., Chen, H. B., Tong, Q., Li, X. Y., Zhu, M. J., Wu, Z. F., Zhou, R. and Zhao, S. H. (2011). Molecular characterization of Caveolin-1 in pigs infected with Haemophilus parasuis, *J. Immunol.* 186(5), 3031-3046.

Lordelo, M. M., Gaspar, A. M., Le Bellego, L. and Freire, J. P. B. (2008). Isoleucine and valine supplementation of a low-protein corn-wheat-soybean meal-based diet for piglets: growth performance and nitrogen balance, *J. Anim. Sci.* 86(11), 2936-2941.

Luise, D., Lauridsen, C., Bosi, P. and Trevisi, P. (2019a). Methodology and application of Escherichia coli F4 and F18 encoding infection models in post-weaning pigs, *J. Anim. Sci. Biotechnol.* 10(1), 53.

Luise, D., Motta, V., Bertocchi, M., Salvarani, C., Clavenzani, P., Fanelli, F., Pagotto, U., Bosi, P. and Trevisi, P. (2019b). Effect of Mucine 4 and fucosyltransferase 1 genetic variants on gut homoeostasis of growing healthy pigs, *J. Anim. Physiol. Anim. Nutr. (Berl)* 103(3), 801-812.

Lundeheim, N. (1979). Genetic analysis of respiratory diseases in pigs, *Acta Agric. Scand.* 29(3), 209-215.

Luppi, A. (2017). Swine enteric colibacillosis: diagnosis, therapy and antimicrobial resistance, *Porcine Health Manag.* 3(1), 16.

Luppi, A., Gibellini, M. V., Gin, T., Vangroenweghe, F., Vandenbroucke, V., Bauerfeind, R., Bonilauri, P., Labarque, G. and Hidalgo, Á (2016). Prevalence of virulence factors in enterotoxigenic *Escherichia coli* isolated from pigs with post-weaning diarrhoea in Europe, *Porcine Health Manag.* 2, 20.

Mäkivuokko, H., Lahtinen, S. J., Wacklin, P., Tuovinen, E., Tenkanen, H., Nikkilä, J., Björklund, M., Aranko, K., Ouwehand, A. C. and Mättö, J. (2012). Association between the ABO blood group and the human intestinal microbiota composition, *BMC Microbiol.* 12, 94.

Mantle, M. and Allen, A. (1981). Isolation and characterization of the native glycoprotein from pig small-intestinal mucus, *Biochem. J.* 195(1), 267-275.

Marsteller, T. A., Armbruster, G., Bane, D. P., Gebhart, C. J., Weatherford, J. and Thacker, B. (2003). Monitoring the prevalence of Lawsonia intracellularis IgG antibodies using serial sampling in growing breeding swine herds, *J. Swine Health Prod.* 11, 127-130.

Massacci, F. R., Tofani, S., Forte, C., Bertocchi, M., Lovito, C., Orsini, S., Tentellini, M., Marchi, L., Lemonnier, G., Luise, D., Blanc, F., Castinel, A., Bevilacqua, C., Rogel-Gaillard, C., Pezzotti, G., Estellé, J., Trevisi, P. and Magistrali, C. F. (2020). Host genotype and amoxicillin administration affect the incidence of diarrhoea and faecal microbiota of weaned piglets during a natural multiresistant ETEC infection, *J. Anim. Breed. Genet.* 137(1), 60-72.

McDonald, D. E., Pethick, D. W., Mullan, B. P. and Hampson, D. J. (2001). Increasing viscosity of the intestinal contents alters small intestinal structure and intestinal growth, and stimulate proliferation of enterotoxigenic *Escherichia coli* in newly-weaned pigs, *Br. J. Nutr.* 86(4), 487-498.

McIntyre, N., Smith, D. G. E., Shaw, D. J., Thomson, J. R. and Rhind, S. M. (2003). Immunopathogenesis of experimentally induced proliferative enteropathy in pigs, *Vet. Pathol.* 40(4), 421-432.

McOrist, S. and Gebhart, C. J. (1999). Porcine proliferative entero- pathies. In: Straw, B., Mengeling, W., D'Allaire, S. and Taylor, D. (Eds), *Diseases of Swine* (8th edn.), Ames, IA: Iowa State University Press, pp. 521-534.

Meijerink, E., Neuenschwander, S., Fries, R., Dinter, A., Bertschinger, H. U., Stranzinger, G. and Vögeli, P. (2000). A DNA polymorphism influencing alpha(1,2)fucosyltransferase activity of the pig FUT1 enzyme determines susceptibility of small intestinal epithelium to Escherichia coli F18 adhesion, *Immunogenetics* 52(1-2), 129-136.

Mendis, M., Leclerc, E. and Simsek, S. (2016). Arabinoxylans, gut microbiota and immunity, *Carbohydr. Polym.* 139, 159-166.

Mikkelsen, L. L., Naughton, P. J., Hedemann, M. S. and Jensen, B. B. (2004). Effects of physical properties of feed on microbial ecology and survival of *Salmonella enterica* serovar Typhimurium in the pig gastrointestinal tract, *Appl. Environ. Microbiol.* 70(6), 3485-3492.

Millet, S., Kumar, S., De Boever, J., Ducatelle, R. and De Brabander, D. (2012). Effect of feed processing on growth performance and gastric mucosa integrity in pigs from weaning until slaughter, *Anim. Feed Sci. Technol.* 175(3-4), 175-181.

Millet, S., van Hees, H., Janssens, G. P. J. and De Smet, S. (2019). The effect of an 18-hour delay in solid feed provisioning on the feed intake and performance of piglets in the first weeks after weaning, *Livest. Sci.* 228, 49-52.

Modesto, M., Stefanini, I., D'Aimmo, M. R., Mazzoni, M., Trevisi, P., Tittarelli, C., Bosi, P. and Biavati, B. (2007). Effect of probiotic inocula on the population density of lactic acid bacteria and enteric pathogens in the intestine of weaning piglets. In: 3rd QLIF Congress, Hohenheim, Germany, March 20-23, 2007.

Modesto, M., D'Aimmo, M. R., Stefanini, I., Trevisi, P., De Filippi, S., Casini, L., Mazzoni, M., Bosi, P. and Biavati, B. (2009). A novel strategy to select Bifidobacterium strains and prebiotics as natural growth promoters in newly weaned pigs, *Livest. Sci.* 122(2-3), 248-258.

Mølbak, L., Johnsen, K., Boye, M., Jensen, T. K., Johansen, M., Møller, K. and Leser, T. D. (2008). The microbiota of pigs influenced by diet texture and severity of Lawsonia intracellularis infection, *Vet. Microbiol.* 128(1-2), 96-107.

Molist, F., de Segura, A. G., Gasa, J., Hermes, R. G., Manzanilla, E. G., Anguita, M. and Pérez, J. F. (2009). Effects of the insoluble and soluble dietary fibre on the physicochemical properties of digesta and the microbial activity in early weaned piglets, *J. Anim. Sci. Technol.* 149(3-4), 346-353.

Molist, F., Van Oostrum, M., Pérez, J. F., Mateos, G. G., Nyachoti, C. M. and Van Der Aar, P. J. (2014). Relevance of functional properties of dietary fibre in diets for weanling pigs, *J. Anim. Sci. Technol.* 189, 1-10.

Montagne, L., Pluske, J. R. and Hampson, D. J. (2003). A review of interactions between dietary fibre and the intestinal mucosa, and their consequences on digestive health in young non-ruminant animals, *Anim. Feed Sci. Technol.* 108(1-4), 95-117.

Mößeler, A., Köttendorf, S., Große Liesner, V. and Kamphues, J. (2010). Impact of diets' physical form (particle size; meal/pelleted) on the stomach content (dry matter content, pH, chloride concentration) of pigs, *Livest. Sci.* 134(1-3), 146-148.

Motta, V., Trevisi, P., Bertolini, F., Ribani, A., Schiavo, G., Fontanesi, L. and Bosi, P. (2017). Exploring gastric bacterial community in young pigs, *PLoS ONE* 12(3), e0173029.

Motta, V., Luise, D., Bosi, P. and Trevisi, P. (2019). Faecal microbiota shift during weaning transition in piglets and evaluation of AO blood types as shaping factor for the bacterial community profile, *PLoS ONE* 14, 1-18.

Munasinghe, L. L., Robinson, J. L., Harding, S. V., Brunton, J. A. and Bertolo, R. F. (2017). Protein synthesis in mucin-producing tissues is conserved when dietary threonine is limiting in piglets, *J. Nutr.* 147(2), 202-210.

Nabuurs, M. J. A., Hoogendoorn, A. and Van Zijderveld-Van Bemmel, A. (1996). Effect of supplementary feeding during the sucking period on net absorption from the small intestine of weaned pigs, *Res. Vet. Sci.* 61(1), 72-77.

Nagy, B. and Fekete, P. Z. (2005). Enterotoxigenic *Escherichia coli* in veterinary medicine, *Int. J. Med. Microbiol.* 295(6-7), 443-454.

Naresh, R. and Hampson, D. J. (2010). Attraction of *Brachyspira pilosicoli* to mucin, *Microbiology* 156(1), 191-197.

Nguyen, T. V., Yuan, L., Azevedo, M. S., Jeong, K. I., Gonzalez, A. M., Iosef, C., Lovgren-Bengtsson, K., Morein, B., Lewis, P. and Saif, L. J. (2006). High titers of circulating maternal antibodies suppress effector and memory B-cell responses induced by an attenuated rotavirus priming and rotavirus-like particle-immunostimulating complex boosting vaccine regimen, *Clin. Vaccine Immunol.* 13(4), 475-485.

Nyachoti, C. M., Omogbenigun, F. O., Rademacher, M. and Blank, G. (2006). Performance responses and indicators of gastrointestinal health in early-weaned pigs fed low-protein amino acid-supplemented diets, *J. Anim. Sci.* 84(1), 125-134.

Opapeju, F. O., Rademacher, M., Blank, G. and Nyachoti, C. M. (2008). Effect of low-protein amino acid-supplemented diets on the growth performance, gut morphology, organ weights and digesta characteristics of weaned pigs, *Animal* 2(10), 1457-1464.

Opapeju, F. O., Krause, D. O., Payne, R. L., Rademacher, M. and Nyachoti, C. M. (2009). Effect of dietary protein level on growth performance, indicators of enteric health, and gastrointestinal microbial ecology of weaned pigs induced with postweaning colibacillosis, *J. Anim. Sci.* 87(8), 2635-2643.

Opapeju, F. O., Rodriguez-Lecompte, J. C., Rademacher, M., Krause, D. O. and Nyachoti, C. M. (2015). Les diètes faibles en protéines brutes modulent la réponse intestinale chez les porcs sevrés exposés à *Escherichia coli* K88, *Can. J. Anim. Sci.* 95(1), 71-78.

Opriessnig, T., Karuppannan, A. K., Beckler, D., Ait-Ali, T., Cubas-Atienzar, A. and Halbur, P. G. (2019). Bacillus pumilus probiotic feed supplementation mitigates *Lawsonia intracellularis* shedding and lesions, *Vet. Res.* 50(1), 85.

Ouyang, J., Zeng, W., Ren, J., Yan, X., Zhang, Z., Yang, M., Han, P., Huang, X., Ai, H. and Huang, L. (2012). Association of B3GNT5 polymorphisms with susceptibility to ETEC F4ab/ac in the White Duroc × Erhualian intercross and 15 outbred pig breeds, *Biochem. Genet.* 50(1-2), 19-33.

Oxberry, S. L. and Hampson, D. J. (2003). Epidemiological studies of Brachyspira pilosicoli in two Australian piggeries, *Vet. Microbiol.* 93(2), 109-120.

Palombo, V., Loor, J. J., D'Andrea, M., Vailati-Riboni, M., Shahzad, K., Krogh, U. and Theil, P. K. (2018). Transcriptional profiling of swine mammary gland during the transition from colostrogenesis to lactogenesis using RNA sequencing, *BMC Genomics* 19(1), 322.

Partanen, K. H. and Mroz, Z. (1999). Organic acids for performance enhancement in pig diets, *Nutr. Res. Rev.* 12(1), 117-145.

Pearce, S. C., Sanz-Fernandez, M. V., Hollis, J. H., Baumgard, L. H. and Gabler, N. K. (2014). Short-term exposure to heat stress attenuates appetite and intestinal integrity in growing pigs, *J. Anim. Sci.* 92(12), 5444-5454.

Peng, X., Hu, L., Liu, Y., Yan, C., Fang, Z. F., Lin, Y., Xu, S. Y., Li, J., Wu, C. M., Chen, D. W., Sun, H., Wu, D. and Che, L. Q. (2016). Effects of low-protein diets supplemented with indispensable amino acids on growth performance, intestinal morphology and immunological parameters in 13 to 35 kg pigs, *Animal* 10(11), 1812-1820.

Piccolo, B. D., Mercer, K. E., Bhattacharyya, S., Bowlin, A. K., Saraf, M. K., Pack, L., Chintapalli, S. V., Shankar, K., Adams, S. H., Badger, T. M. and Yeruva, L. (2017). Early postnatal diets affect the bioregional small intestine microbiome and ileal metabolome in neonatal pigs, *J. Nutr.* 147(8), 1499-1509.

Pieper, R., Kröger, S., Richter, J. F., Wang, J., Martin, L., Bindelle, J., Htoo, J. K., von Smolinski, D., Vahjen, W., Zentek, J. and Van Kessel, A. G. (2012). Fermentable fiber ameliorates fermentable protein-induced changes in microbial ecology, but not the mucosal response, in the colon of piglets, *J. Nutr.* 142(4), 661-667.

Pieper, R., Villodre Tudela, C., Taciak, M., Bindelle, J., Pérez, J. F. and Zentek, J. (2016). Health relevance of intestinal protein fermentation in young pigs, *Anim. Heal. Res. Rev.* 17(2), 137-147.

Poulsen, A. R., Luise, D., Curtasu, M. V., Sugiharto, S., Canibe, N., Trevisi, P. and Lauridsen, C. (2018). Effects of alpha-(1,2)-fucosyltransferase genotype variants on plasma metabolome, immune responses and gastrointestinal bacterial enumeration of pigs pre- and post- weaning, *PLoS ONE*, 13(8), e0202970.

Prims, S., Tambuyzer, B., Vergauwen, H., Huygelen, V., Van Cruchten, S. V., Ginneken, C. V. and Casteleyn, C. (2016). Intestinal immune cell quantification and gram type classification of the adherent microbiota in conventionally and artificially reared, normal and low birth weight piglets, *Livest. Sci.* 185, 1-7.

Prims, S., Pintens, N., Vergauwen, H., Van Cruchten, S., Van Ginneken, C. and Casteleyn, C. (2017). Effect of artificial rearing of piglets on the volume densities of M cells in the tonsils of the soft palate and ileal Peyer's patches, *Vet. Immunol. Immunopathol.* 184, 1-7.

Prims, S., Jurgens, B., Hole, C. V., Van Cruchten, S., Van Ginneken, C. and Casteleyn, C. (2018). The porcine tonsils and Peyer's patches: a stereological morphometric analysis in conventionally and artificially reared piglets, *Vet. Immunol. Immunopathol.* 206, 9-15.

Priori, D., Colombo, M., Koopmans, S. J., Jansman, A. J. M., van der Meulen, J., Trevisi, P. and Bosi, P. (2016). The A0 blood group genotype modifies the jejunal glycomic binding pattern profile of piglets early associated with a simple or complex microbiota, *J. Anim. Sci.* 94(2), 592-601.

Rao, A., Jump, R. L., Pultz, N. J., Pultz, M. J. and Donskey, C. J. (2006). In vitro killing of nosocomial pathogens by acid and acidified nitrite, *Antimicrob. Agents Chemother.* 50(11), 3901-3904.

Ren, J., Yan, X., Ai, H., Zhang, Z., Huang, X., Ouyang, J., Yang, M., Yang, H., Han, P., Zeng, W., Chen, Y., Guo, Y., Xiao, S., Ding, N. and Huang, L. (2012). Susceptibility towards enterotoxigenic *Escherichia coli* F4ac diarrhea Is Governed by the MUC13 Gene in Pigs, *PLoS ONE*, 7(9), e44573.

Ren, M., Cai, S., Zhou, T., Zhang, S., Li, S., Jin, E., Che, C., Zeng, X., Zhang, T. and Qiao, S. (2019). Isoleucine attenuates infection induced by E. coli challenge through the modulation of intestinal endogenous antimicrobial peptide expression and the inhibition of the increase in plasma endotoxin and IL-6 in weaned pigs, *Food Funct.* 10(6), 3535-3542.

Ren, M., Zhang, S. H., Zeng, X. F., Liu, H. and Qiao, S. Y. (2015). Branched-chain amino acids are beneficial to maintain growth performance and intestinal immune-related function in weaned piglets fed protein restricted diet, *Asian-Australas. J. Anim. Sci.* 28(12), 1742-1750.

Richter, J. F., Pieper, R., Zakrzewski, S. S., Günzel, D., Schulzke, J. D. and Van Kessel, A. G. (2014). Diets high in fermentable protein and fibre alter tight junction protein composition with minor effects on barrier function in piglet colon, *Br. J. Nutr.* 111(6), 1040-1049.

Rieger, J., Janczyk, P., Hünigen, H., Neumann, K. and Plendl, J. (2015). Intraepithelial lymphocyte numbers and histomorphological parameters in the porcine gut after *Enterococcus faecium* NCIMB 10415 feeding in a Salmonella Typhimurium challenge, *Vet. Immunol. Immunopathol.* 164(1-2), 40-50.

Rong, J., Zhang, W., Wang, X., Fan, H., Lu, C. and Yao, H. (2012). Identification of candidate susceptibility and resistance genes of mice infected with streptococcus suis type 2', *PLoS ONE*, 7(2), 2.

Roof, M. B., Roth, J. and Kramer, T. T. (1992). Porcine salmonellosis: characterization, immunity, and potential vaccines. *Compend. Cont. Educ, Pract. Vet.* 14, 411-423.

Roselli, M., Finamore, A., Garaguso, I., Britti, M. S. and Mengheri, E. (2003). Zinc oxide protects cultured enterocytes from the damage induced by *Escherichia coli*, *J. Nutr.* 133(12), 4077-4082.

Ross, J. W., Hale, B. J., Gabler, N. K., Rhoads, R. P., Keating, A. F. and Baumgard, L. H. (2015). Physiological consequences of heat stress in pigs, *Anim. Prod. Sci.* 55(12), 1381–1390.

Ross, J. W., Hale, B. J., Seibert, J. T., Romoser, M. R., Adur, M. K., Keating, A. F. and Baumgard, L. H. (2017). Physiological mechanisms through which heat stress compromises reproduction in pigs, *Mol. Reprod. Dev.* 84(9), 934–945.

Rothkötter, H., Möllhoff, S. and Pabst, R. (1999). The influence of age and breeding conditions on the number and proliferation of intraepithelial lymphocytes in pigs, *Scand. J. Immunol.* 50(1), 31–38.

Rowland, A. and Lawson, G. H. K. (1992). Porcine proliferative enteropathies. In: Straw, B., Mengeling, W., D'Allaire, S. and Taylor, D. (Eds), *Diseases of Swine* (7th edn.), Ames, IA: Iowa State University Press, pp. 560–569.

Sako, F., Gasa, S., Makita, A., Hayashi, A. and Nozawa, S. (1990). Human blood group glycosphingolipids of porcine erythrocytes, *Arch. Biochem. Biophys.* 278(1), 228–237.

Salcedo, J., Frese, S. A., Mills, D. A.. and Barile, D. (2016). Characterization of porcine milk oligosaccharides during early lactation and their relation to the fecal microbiome, *J. Dairy Sci.* 99(10), 7733–7743.

Saraf, M. K., Piccolo, B. D., Bowlin, A. K., Mercer, K. E., LeRoith, T., Chintapalli, S. V., Shankar, K., Badger, T. M. and Yeruva, L. (2017). Formula diet driven microbiota shifts tryptophan metabolism from serotonin to tryptamine in neonatal porcine colon, *Microbiome* 5(1), 77.

Scharek-Tedin, L., Pieper, R., Vahjen, W., Tedin, K., Neumann, K. and Zentek, J. (2013). Bacillus cereus var. Toyoi modulates the immune reaction and reduces the occurrence of diarrhea in piglets challenged with Salmonella Typhimurium DT104, *J. Anim. Sci.* 91(12), 5696–5704.

Siepert, B., Reinhardt, N., Kreuzer, S., Bondzio, A., Twardziok, S., Brockmann, G., Nöckler, K., Szabó, I., Janczyk, P., Pieper, R. and Tedin, K. (2014). Enterococcus faecium NCIMB 10415 supplementation affects intestinal immune-associated gene expression in post-weaning piglets, *Vet. Immunol. Immunopathol.* 157(1-2), 65–77.

Sørensen, M. T., Vestergaard, E. M., Jensen, S. K., Lauridsen, C. and Højsgaard, S. (2009). Performance and diarrhoea in piglets following weaning at seven weeks of age: challenge with E. coli O 149 and effect of dietary factors, *Livest. Sci.* 123(2-3), 314–321.

Spreeuwenberg, M. A. M., Verdonk, J. M. A. J., Gaskins, H. R. and Verstegen, M. W. A. (2001). Small intestine epithelial barrier function is compromised in pigs with low feed intake at weaning, *J. Nutr.* 131(5), 1520–1527.

Stege, H., Jensen, T. K., Møller, K., Vestergaard, K., Baekbo, P. and Jorsal, S. E. (2004). Infection dynamics of Lawsonia intracellularis in pig herds, *Vet. Microbiol.* 104(3-4), 197–206.

Stumpff, F., Lodemann, U., Van Kessel, A. G., Pieper, R., Klingspor, S., Wolf, K., Martens, H., Zenteck, J. and Aschenbach, J. R. (2013). Effects of dietary fibre and protein on urea transport across the cecal mucosa of piglets. *J. Comp. Physiol. B* 183(8), 1053–1063.

Sulabo, R. C., Tokach, M. D., Dritz, S. S., Goodband, R. D., DeRouchey, J. M. and Nelssen, J. L. (2010). Effects of varying creep feeding duration on the proportion of pigs consuming creep feed and neonatal pig performance, *J. Anim. Sci.* 88(9), 3154–3162.

Taube, V. A., Neu, M. E., Hassan, Y., Verspohl, J., Beyer-Bach, M. and Kamphues, J. (2009). Effects of dietary addi- tives (potassium diformate/organic acids) as well as influences of grinding intensity (coarse/fine) of diets for weaned piglets experimentally infected with Salmonella Derby or Escherichia coli, *J. Anim. Physiol. Anim. Nutr.* 93(3), 350–358.

Thomsen, L. E., Bach Knudsen, K. E., Jensen, T. K., Christensen, A. S., Møller, K. and Roepstorff, A. (2007). The effect of fermentable carbohydrates on experimental swine dysentery and whip worm infections in pigs, *Vet. Microbiol.* 119(2-4), 152–163.

Torrallardona, D., Conde, R., Badiola, I. and Polo, J. (2007). Evaluation of spray dried animal plasma and calcium formate as alternatives to colistin in piglets experimentally infected with Escherichia coli K99, *Livest. Sci.* 108(1-3), 303–306.

Trevisi, P., Priori, D., Gandolfi, G., Colombo, M., Coloretti, F., Goossens, T. and Bosi, P. (2012). In vitro test on the ability of a yeast cell wall based product to inhibit the *Escherichia coli* F4ac adhesion on the brush border of porcine intestinal villi, *J. Anim. Sci.* 90 (Suppl. 4), 275–277.

Trevisi, P., Colombo, M., Priori, D., Fontanesi, L., Galimberti, G., Calò, G., Motta, V., Latorre, R., Fanelli, F., Mezzullo, M., Pagotto, U., Gherpelli, Y., D'Inca, R. and Bosi, P. (2015a). Comparison of three patterns of feed supplementation with live Saccharomyces cerevisiae yeast on post-weaning diarrhea, health status, and blood metabolic profile of susceptible weaning pigs orally challenged with Escherichia coli F4ac, *J. Anim. Sci.* 93(5), 2225–2233.

Trevisi, P., Corrent, E., Mazzoni, M., Messori, S., Priori, D., Gherpelli, Y., Simongiovanni, A. and Bosi, P. (2015b). Effect of added dietary threonine on growth performance, health, immunity and gastrointestinal function of weaning pigs with differing genetic susceptibility to *Escherichia coli* infection and challenged with E. coli K88ac, *J. Anim. Physiol. Anim. Nutr. (Berl)* 99(3), 511–520.

Trevisi, P., Latorre, R., Priori, D., Luise, D., Archetti, I., Mazzoni, M., D'Inca, R. and Bosi, P. (2017a). Effect of feed supplementation with live yeast on the intestinal transcriptome profile of weaning pigs orally challenged with *Escherichia coli* F4, *Animal* 11(1), 33–44.

Trevisi, P., Miller, B., Patel, D., Bolognesi, A., Bortolotti, M. and Bosi, P. (2017b). Two different in vitro tests confirm the blocking activity of d-galactose lectins on the adhesion of Escherichia coli F4 to pig brush border receptors, *Ital. J. Anim. Sci.* 16(1), 101–107.

Trevisi, P., Luise, D., Won, S., Slacedo, J., Bertocchi, M., Barile, D. and Bosi, P. (2020). Variations in porcine colostrum oligosaccharide composition between breeds and in association with sow maternal performance, *J. Anim. Sci. Biotechnol.* 11, 21.

Trott, D. J., Huxtable, C. R. and Hampson, D. J. (1996). Experimental infection of newly weaned pigs with human and porcine strains of *Serpulina pilosicoli*, *Infect. Immun.* 64(11), 4648–4654.

Tudela, C. V., Boudry, C., Stumpff, F., Aschenbach, J. R., Vahjen, W., Zentek, J. and Pieper, R. (2015). Down-regulation of monocarboxylate transporter 1 (MCT1) gene expression in the colon of piglets is linked to bacterial protein fermentation and pro-inflammatory cytokine-mediated signalling, *Br. J. Nutr.* 113(4), 610–617.

Uthe, J. J., Qu, L., Couture, O., Bearson, S. M. D., O'Connor, A. M., McKean, J. D., Torres, Y. R., Dekkers, J. C., Nettleton, D. and Tuggle, C. K. (2011). Use of bioinformatic SNP predictions in differentially expressed genes to find SNPs associated with Salmonella colonization in swine, *J. Anim. Breed. Genet.* 128(5), 354–365.

Van der Heijden, H. M. J. F., Bakker, J., Elbers, A. R. W., Vos, J. H., Weyns, A., de Smet, M. and McOrist, S. (2004). Prevalence of exposure and infection of *Lawsonia intracellularis* among slaughter-age pigs, *Res. Vet. Sci.* 77(3), 197–202.

Verhelst, R., Schroyen, M., Buys, N. and Niewold, T. (2010). The effects of plant polyphenols on enterotoxigenic Escherichia coli adhesion and toxin binding, *Lives* 133(1-3), 101-103.

Verhelst, R., Schroyen, M., Buys, N. and Niewold, T. (2014). Dietary polyphenols reduce diarrhea in enterotoxigenic *Escherichia coli* (ETEC) infected post-weaning piglets, *Livest. Sci.* 160, 138-140.

Visscher, C., Kruse, A., Sander, S., Keller, C., Mischok, J., Tabeling, R., Henne, H., Deitmer, R. and Kamphues, J. (2018). Experimental studies on effects of diet on *Lawsonia intracellularis* infections in fattening boars in a natural infection modelì, *Acta Vet. Scand.* 60(1), 22.

Walczak, K. (2015). Insight into the epidemiology of swine dysentery by analyzing treatment records and using simulation modelling. Master of Science Dissertation, University of Guelph, Guelph, Canada.

Wan, K., Li, Y., Sun, W., An, R., Tang, Z., Wu, L., Chen, H. and Sun, Z. (2020). Effects of dietary calcium pyruvate on gastrointestinal tract development, intestinal health and growth performance of newly weaned piglets fed low-protein diets, *J. Appl. Microbiol.* 128(2), 355-365.

Wang, J., Zeng, Y., Wang, S., Liu, H., Zhang, D., Zhang, W., Wang, Y. and Ji, H. (2018). Swine-derived probiotic Lactobacillus plantarum inhibits growth and adhesion of enterotoxigenic *Escherichia coli* and mediates host defense, *Front. Microbiol.* 9, 1364.

Wang, J., Zhang, W., Wang, S., Liu, H., Zhang, D., Wang, Y. and Ji, H. (2019). Swine-derived probiotic lactobacillus plantarum modulates porcine intestinal endogenous host defense peptide synthesis through tlr2/mapk/ap-1 signaling pathway, *Front. Immunol.* 10, 2691.

Wang, S. J., Liu, W. J., Yang, L. G., Sargent, C. A., Liu, H. B., Wang, C., Liu, X. D., Zhao, S. H., Affara, N. A., Liang, A. X. and Zhang, S. J. (2012). Effects of FUT1 gene mutation on resistance to infectious disease, *Mol. Biol. Rep.* 39(3), 2805-2810.

Wang, Y., Ren, J., Lan, L., Yan, X., Huang, X., Peng, Q., Tang, H., Zhang, B., Ji, H. and Huang, L. (2007). Characterization of polymorphisms of transferrin receptor and their association with susceptibility to ETEC F4ab/ac in pigs, *J. Anim. Breed. Genet.* 124(4), 225-229.

Wellington, M. O., Hamonic, K., Krone, J. E. C., Htoo, J. K., Van Kessel, A. G. and Columbus, D. A. (2020). Effect of dietary fiber and threonine content on intestinal barrier function in pigs challenged with either systemic *E. coli* lipopolysaccharide or enteric Salmonella typhimurium, *J. Anim. Sci. Biotechnol.* 11, 1-12.

Wellock, I. J., Fortomaris, P. D., Houdijk, J. G. M. and Kyriazakis, I. (2008). Effects of dietary protein supply, weaning age and experimental enterotoxigenic Escherichia coli infection on newly weaned pigs, *Health. Animal* 2(6), 834-842.

Wen, X., Wang, L., Zheng, C., Yang, X., Ma, X., Wu, Y., Chen, Z. and Jiang, Z. (2018). Fecal scores and microbial metabolites in weaned piglets fed different protein sources and levels, *Anim. Nutr.* 4(1), 31-36.

Wilcock, B. P. and Olander, H. J. (1977). The pathogenesis of porcine rectal stricture. II. Experimental salmonellosis and ischemic proctitis, *Vet. Pathol.* 14(1), 43-55.

Wilcock, B. P. and Schwartz, K. J. (1992). Salmonellosis. In: Leman, A. D., Straw, B. E., Mengeling, W. L., D'Allaire, S. and Taylor, D. J. (Eds), *Diseases of Swine* (7th edn.), Ames, IA: Iowa State University Press, pp.57-583.

Wills, R. W. (2000). Diarrhea in growing-finishing swine, *Vet. Clin. North Am. Food Anim. Pract.* 16(1), 135-161.

Wu, Y., Jiang, Z., Zheng, C., Wang, L., Zhu, C., Yang, X., Wen, X. and Ma, X. (2015). Effects of protein sources and levels in antibiotic-free diets on diarrhea, intestinal morphology, and expression of tight junctions in weaned piglets, *Anim. Nutr.* 1(3), 170–176.

Xia, S., Yao, W., Zou, B., Lu, Y., Lu, N., Lei, H. and Xia, D. (2016). Effects of potassium diformate on the gastric function of weaning piglets, *Anim. Prod. Sci.* 56(7), 1161–1166.

Yang, Q., Huang, X., Wang, P., Yan, Z., Sun, W., Zhao, S. and Gun, S. (2019). Longitudinal development of the gut microbiota in healthy and diarrheic piglets induced by age-related dietary changes, *MicrobiologyOpen* 8(12), e923.

Yeruva, L., Spencer, N. E., Saraf, M. K., Hennings, L., Bowlin, A. K., Cleves, M. A., Mercer, K., Chintapalli, S. V., Shankar, K., Rank, R. G., Badger, T. M. and Ronis, M. J. (2016). Formula diet alters small intestine morphology, microbial abundance and reduces VE-cadherin and IL-10 expression in neonatal porcine model, *BMC Gastroenterol.* 16(1), 40.

Yi, G. F., Carroll, J. A., Allee, G. L., Gaines, A. M., Kendall, D. C., Usry, J. L., Toride, Y. and Izuru, S. (2005). Effect of glutamine and spray-dried plasma on growth performance, small intestinal morphology, and immune responses of Escherichia coli K88+-challenged weaned pigs, *J. Anim. Sci.* 83(3), 634–643.

Yoo, S. S., Field, C. J. and McBurney, M. I. (1997). Glutamine supplementation maintains intramuscular glutamine concentrations and normalizes lymphocyte function in infected early weaned pigs, *J. Nutr.* 127(11), 2253–2259.

Yu, D., Zhu, W. and Hang, S. (2019a). Effects of long-term dietary protein restriction on intestinal morphology, digestive enzymes, gut hormones, and colonic microbiota in pigs, *Animals (Basel)* 9(4), 1–14.

Yu, D., Zhu, W. and Hang, S. (2019b). Effects of low-protein diet on the intestinal morphology, digestive enzyme activity, blood urea nitrogen, and gut microbiota and metabolites in weaned pigs, *Arch. Anim. Nutr.* 73(4), 287–305.

Yu, T., Zhu, C., Chen, S., Gao, L., Lv, H., Feng, R., Zhu, Q., Xu, J., Chen, Z. and Jiang, Z. (2017). Dietary high zinc oxide modulates the microbiome of ileum and colon in weaned piglets, *Front. Microbiol.* 8, 825.

Zhang, B., Ren, J., Yan, X., Huang, X., Ji, H., Peng, Q., Zhang, Z. and Huang, L. (2008). Investigation of the porcine MUC13 gene: isolation, expression, polymorphisms and strong association with susceptibility to enterotoxigenic *Escherichia coli* F4ab/ac, *Anim. Genet.* 39(3), 258–266.

Zhang, L., Mu, C., He, X., Su, Y., Mao, S., Zhan, J., Hauke, S. and Zhu, W. (2016). Effects of dietary fibre source on microbiota composition in the large intestine of suckling piglets, *FEMS Microbiol. Lett.* 363(14), fnw138.

Zhang, Z., Kwawukume, A., Moossavi, S., Sepehri, S., Nyachoti, M. and Khafipour, E. (2018). Time series and correlation network analyses to identify the role of maternal microbiomes on development of piglet gut microbiome and susceptibility to neonatal porcine diarrhea, *J. Anim. Sci.* 96(Suppl. 2), 213.

Zhao, S., Zhu, M. and Chen, H. (2012). Immunogenomics for identification of disease resistance genes in pigs: a review focusing on Gram-negative bacilli. *J. Anim. Sci. Biotechnol.* 3, 34. doi: 10.1186/2049-1891-3-34.

Zhou, X., Zhang, Y., Wu, X., Wan, D. and Yin, Y. (2018). Effects of dietary serine supplementation on intestinal integrity, inflammation and oxidative status in early-weaned piglets, *Cell. Physiol. Biochem.* 48(3), 993–1002.

www.ingramcontent.com/pod-product-compliance
Lightning Source LLC
Chambersburg PA
CBHW050526270326
41926CB00015B/3091